LIFE CYCLE ASSET MANAGEMENT OF URANIUM MINING AND PROCESSING FACILITIES

The following States are Members of the International Atomic Energy Agency:

AFGHANISTAN
ALBANIA
ALGERIA
ANGOLA
ANTIGUA AND BARBUDA
ARGENTINA
ARMENIA
AUSTRALIA
AUSTRIA
AZERBAIJAN
BAHAMAS
BAHRAIN
BANGLADESH
BARBADOS
BELARUS
BELGIUM
BELIZE
BENIN
BOLIVIA, PLURINATIONAL STATE OF
BOSNIA AND HERZEGOVINA
BOTSWANA
BRAZIL
BRUNEI DARUSSALAM
BULGARIA
BURKINA FASO
BURUNDI
CABO VERDE
CAMBODIA
CAMEROON
CANADA
CENTRAL AFRICAN REPUBLIC
CHAD
CHILE
CHINA
COLOMBIA
COMOROS
CONGO
COOK ISLANDS
COSTA RICA
CÔTE D'IVOIRE
CROATIA
CUBA
CYPRUS
CZECH REPUBLIC
DEMOCRATIC REPUBLIC OF THE CONGO
DENMARK
DJIBOUTI
DOMINICA
DOMINICAN REPUBLIC
ECUADOR
EGYPT
EL SALVADOR
ERITREA
ESTONIA
ESWATINI
ETHIOPIA
FIJI
FINLAND
FRANCE
GABON
GAMBIA, THE
GEORGIA
GERMANY
GHANA
GREECE
GRENADA
GUATEMALA
GUINEA
GUYANA
HAITI
HOLY SEE
HONDURAS
HUNGARY
ICELAND
INDIA
INDONESIA
IRAN, ISLAMIC REPUBLIC OF
IRAQ
IRELAND
ISRAEL
ITALY
JAMAICA
JAPAN
JORDAN
KAZAKHSTAN
KENYA
KOREA, REPUBLIC OF
KUWAIT
KYRGYZSTAN
LAO PEOPLE'S DEMOCRATIC REPUBLIC
LATVIA
LEBANON
LESOTHO
LIBERIA
LIBYA
LIECHTENSTEIN
LITHUANIA
LUXEMBOURG
MADAGASCAR
MALAWI
MALAYSIA
MALI
MALTA
MARSHALL ISLANDS
MAURITANIA
MAURITIUS
MEXICO
MONACO
MONGOLIA
MONTENEGRO
MOROCCO
MOZAMBIQUE
MYANMAR
NAMIBIA
NEPAL
NETHERLANDS, KINGDOM OF THE
NEW ZEALAND
NICARAGUA
NIGER
NIGERIA
NORTH MACEDONIA
NORWAY
OMAN
PAKISTAN
PALAU
PANAMA
PAPUA NEW GUINEA
PARAGUAY
PERU
PHILIPPINES
POLAND
PORTUGAL
QATAR
REPUBLIC OF MOLDOVA
ROMANIA
RUSSIAN FEDERATION
RWANDA
SAINT KITTS AND NEVIS
SAINT LUCIA
SAINT VINCENT AND THE GRENADINES
SAMOA
SAN MARINO
SAUDI ARABIA
SENEGAL
SERBIA
SEYCHELLES
SIERRA LEONE
SINGAPORE
SLOVAKIA
SLOVENIA
SOMALIA
SOUTH AFRICA
SPAIN
SRI LANKA
SUDAN
SWEDEN
SWITZERLAND
SYRIAN ARAB REPUBLIC
TAJIKISTAN
THAILAND
TOGO
TONGA
TRINIDAD AND TOBAGO
TUNISIA
TÜRKİYE
TURKMENISTAN
UGANDA
UKRAINE
UNITED ARAB EMIRATES
UNITED KINGDOM OF GREAT BRITAIN AND NORTHERN IRELAND
UNITED REPUBLIC OF TANZANIA
UNITED STATES OF AMERICA
URUGUAY
UZBEKISTAN
VANUATU
VENEZUELA, BOLIVARIAN REPUBLIC OF
VIET NAM
YEMEN
ZAMBIA
ZIMBABWE

The Agency's Statute was approved on 23 October 1956 by the Conference on the Statute of the IAEA held at United Nations Headquarters, New York; it entered into force on 29 July 1957. The Headquarters of the Agency are situated in Vienna. Its principal objective is "to accelerate and enlarge the contribution of atomic energy to peace, health and prosperity throughout the world".

IAEA-TECDOC-2084

LIFE CYCLE ASSET MANAGEMENT OF URANIUM MINING AND PROCESSING FACILITIES

INTERNATIONAL ATOMIC ENERGY AGENCY
VIENNA, 2025

For further information on this publication, please contact:

Nuclear Fuel Cycle and Materials Section
International Atomic Energy Agency
Vienna International Centre
PO Box 100
1400 Vienna, Austria
Email: Official.Mail@iaea.org

Printed by the IAEA in Austria
March 2025
https://doi.org/10.61092/iaea.21lc-i4zg

IAEA Library Cataloguing in Publication Data

Names: International Atomic Energy Agency.
Title: Life cycle asset management of uranium mining and processing facilities / International Atomic Energy Agency.
Description: Vienna : International Atomic Energy Agency, 2025. | Series: IAEA TECDOC series, ISSN 1011-4289 ; no. 2084 | Includes bibliographical references.
Identifiers: IAEAL 25-01741 | ISBN 978-92-0-105525-5 (paperback : alk. paper) | ISBN 978-92-0-105425-8 (pdf)
Subjects: LCSH: Uranium. | Uranium — Safety measures. | Uranium industry. | Uranium — Management.

FOREWORD

Asset management plays an important role in maintaining the competitiveness of mining and processing facilities in a challenging and changing uranium market. Informed decisions have to be made about maintaining assets for the safe, effective and long term sustainable operation of these facilities.

This publication is intended for IAEA Member States with existing uranium production programmes and those considering introducing or reintroducing uranium mining and production. It provides information on assessment and analysis techniques in decision making for strategic management of assets. The publication addresses aspects of asset management in uranium mines and processing facilities, and presents shared experiences and lessons learned, based on operational experience and knowledge. Although this publication is aimed at promoting asset management for uranium production, the approach presented can also benefit other types of facilities for the extraction and production of other radioactive ores.

The IAEA thanks the experts who contributed to the drafting and review of this publication. The IAEA officers responsible for this publication were B. Moldovan and C. Good of the Division of Nuclear Fuel Cycle and Waste Technology.

CONTENTS

1. INTRODUCTION

1.1. BACKGROUND

A fundamental aspect of a national uranium production programme is the need for safe and reliable production of uranium ore concentrate. Sustainability and public acceptance of the uranium mining and processing industry requires that the infrastructure and equipment of mining and processing facilities is proactively managed throughout all phases of operation, to ensure safety of the public and protection of the environment.

The uranium minerals sector has faced economic challenges due to depressed uranium markets, with uranium mine production levels decreasing and few new uranium production projects or expansions in the past decade [1]. The market conditions have also seen facilities transition into a state of care and maintenance. Ageing facilities and weakened economic conditions require the industry to meet the challenge of improving efficiency of uranium mines and processing facilities, all while maintaining a high level of safety.

Global forecasts indicate an increase in uranium demand in response to growth in nuclear energy supply [1], with new nuclear power plants commencing operations and the introduction of new nuclear power technologies into the market. To meet the needs of the nuclear power industry, uranium providers would need to increase production in response. Additionally, uranium supply will require that facilities in care and maintenance come back online, and that planned and prospective mines be realized. Even in favourable market conditions, the management of assets is key to achieving the safe and cost effective operation and maintenance of uranium production facilities.

The sustainability of uranium production requires asset management and maintenance guidelines, strategies, procedures and planning to be in place for uranium mines and processing facilities, throughout all phases of the uranium production cycle.

Life cycle asset management requires uranium production organizations to have engineering and economic tools, as well as risk based decision methods, in the management of currently operating uranium mining and processing facilities, ageing facilities, and facilities that have been in care and maintenance and are likely to resume uranium production. Furthermore, asset management and maintenance programmes need to be considered in the design and planning of new mines and facilities, as well as those transitioning into closure, particularly with regard to the aspects of decontamination and decommissioning.

1.2. OBJECTIVE

The purpose of this publication is to provide good practices in life cycle asset management for safe and sustainable operation of uranium mines and processing facilities. This publication offers general guidance for ageing uranium mines and processing facilities, covering such topics as failure modes, ageing management strategies and related maintenance techniques. It also serves as a common knowledge base on the life cycle asset management of uranium mines and processing facilities, based on industry good practice and international standards. Examples of good industry practices are included in this publication, as well as lessons learned in the operation of ageing uranium mines and processing facilities, to provide practical information on asset maintenance strategies and implementation.

This publication was developed to provide the owner/operators of uranium mines and processing facilities with an overview of good practices for monitoring and maintaining these facilities. Further, this publication provides background and insight that can be utilized by regulatory bodies responsible for the regulation of uranium mines and processing facilities to determine if effective asset management strategies and practices are in place.

1.3. SCOPE

The scope of this publication covers both high level and detailed guidance on life cycle asset management of uranium mines and processing facilities currently in operation. Within this publication, assets are described as equipment or infrastructure required for a mine or processing facility, for the purpose of extraction or processing of uranium ore (e.g. crushers, mills, tanks, structures, haul trucks, etc.). Asset management considerations for new builds, facilities in care and maintenance, revitalization of facilities restarting after care and maintenance, and preparing facilities for closure, are also provided. It includes information on the asset management and maintenance strategies that need to be in place during design, commissioning and early operation stages of facilities, to address potential future ageing issues. In addition, it provides an overview of the topic of, and guidance on, proactive ageing management of uranium mines and processing facilities.

Some of the key aspects included in this publication are:

- Overview of ageing management in uranium mines and processing facilities including the basic concepts of ageing management and the general features of an ageing management programme for these types of facilities.
- Lessons learned on the ageing of structures, systems and components from uranium mines and processing facilities, including industry specific examples.
- Proactive ageing management in uranium mines and processing facilities. Discussion points on this theme include ageing considerations during engineering design, fabrication and construction, operation, care and maintenance, as well as considerations for restart after care and maintenance, and finally prerequisite conditions for decommissioning.
- Elements of an effective ageing management programme including risk based decision processes.
- Degradation mechanisms in uranium mines and processing facilities, including fatigue, corrosion, stress corrosion cracking, flow accelerated corrosion, abrasion and thermal ageing.
- Degradation mitigation in structures, systems and components.
- Methods to assess integrity of structures, systems and components.
- References to international, national and industry standards, codes and guidance materials for asset management of ageing facilities.
- Modern techniques and procedures with respect to ageing management of uranium mines and processing facilities.
- Computerized maintenance management systems (CMMSs).

1.4. STRUCTURE

This publication consists of eight main sections, including the introduction.

Section 2 provides and overview of safety considerations in asset management, introducing bow tie analysis, management of change (MOC) concepts, a structure for risk management and concludes with radiation protection considerations for maintenance workers.

Section 3 provides an overview of life cycle asset management including management systems, elements of an asset management framework, life cycle analysis considerations and an example of an organization structure of an asset management team.

Section 4 provides an introduction to develop an asset management programme, assessment of critical assets and analysis of failure modes, as components of effective management of infrastructure and operating assets. It describes processes for categorizing assets, identifying critical assets, an introduction to the types of maintenance, developing maintenance tactics and spare parts assessment.

Section 5 details the strategies and management considerations for proactive asset management in uranium mines and processing facilities, including defect elimination, work management processes and performance management as part of continuous improvement.

Section 6 provides an overview of degradation mechanisms and conditioning monitoring tactics for infrastructure and uranium mining and processing equipment. Examples of good industry practices provide practical information on implementing asset maintenance strategies and programmes.

Section 7 provides a discussion on key considerations and strategies for planning for facilities that will advance into a state of care and maintenance. This section also details the key considerations and strategies to be considered when restarting a mine or processing facility following a period of care and maintenance.

Section 8 presents the key considerations for asset management for facilities that are entering into closure including the activities of decommissioning and rehabilitation.

The appendices provide supplementary details on life cycle asset management, as presented in Sections 2–8 of this publication. MOC prompts are presented in Appendix II. Appendix III includes a series of equipment failure case studies which bring attention to key planning, maintenance, operation, organization, change management and data management issues, as they relate to asset management within a uranium mine and processing facility. A risk based assessment process is provided in Appendix V. Appendix VIII provides a summary of international and national standards relevant to asset management for further reading.

2. SAFETY CONSIDERATIONS IN ASSET MANAGEMENT STRATEGIES

Asset management strategies play an important role in maintaining the competitiveness of mining and processing facilities in challenging uranium market conditions. Informed, risk based decisions have to be made about maintaining assets for the safe, effective and long term sustainable operation of these facilities. At its foundation, an asset management strategy requires an understanding of the risks associated with the failure of each piece of equipment, and an understanding of how equipment failure scenarios affect the safety and sustainability of uranium mining and processing. In this section of the publication, guidance on assessment and analysis techniques to manage risks are presented, the results of which can be incorporated into the decision making for strategic management of assets.

An introduction to process safety concepts, and its application to asset integrity of uranium mining and processing facilities is provided (Section 2.1). This section includes an overview of the bow tie analysis technique commonly used for process safety risk identification, and how it can be effectively applied to the identification of asset management activities to control process safety risks (Section 2.2). To illustrate these concepts in uranium production facilities, industry examples of process safety bow ties, controls and incident scenarios are presented. Due to uranium processing facilities having a history of catastrophic fires (for those using organic solvents), a specific case study has been included on the risk of fires (Section 2.2.1)

A process is outlined in Section 2.3 for managing assets that are operating under conditions of increased risk.

The establishment of a process to evaluate and control risks that might arise as a result of changes to assets, asset management and/or the asset management system is included in Section 2.4.

In Section 2.5 of the publication a risk assessment process is described, which establishes a basis for decision making associated with risk.

This section concludes with an overview of radiation protection of maintenance workers involved in asset management activities. It provided example of sources of ionizing radiation in uranium mining and processing facilities, general radiation protection measures, and an industry example of a good practice procedure for radiation protection of maintenance workers.

2.1. PROCESS SAFETY

Process safety focusses on designing, operating and maintaining the integrity of facilities that handle hazardous substances, for example those consumed or generated in the processing of uranium ores (e.g. sulphuric acid, ammonia or uranium oxide). The purpose is to manage process hazards in such a way to ensure safe operation of the facility.

Good design principles, engineering and disciplined operating and maintenance practices are essential in managing the integrity of the operating systems and processes that control hazardous substances. An emphasis on both asset integrity and process safety is needed to prevent an unintended primary loss of containment or failure of infrastructure that could result in a major incident. It is important to have multiple layers of control, as illustrated in the bow tie model in Fig. 1, to prevent a process safety incident (this is further discussed in Section 2.2).

The owner/operator of a uranium mine or processing facility needs to determine the organization's resources and allocate accountabilities and responsibilities to ensure the facility is operated in a safe manner [2]. This includes the necessary resources for facility maintenance being defined and allocated to ensure asset integrity.

The owner/operator needs to implement a programme of process safety training and awareness. Generally, this is an ongoing initiative, which commences with the organization's employees onboarding processes, allowing individuals at all levels, including managers and workers, to understand the difference between conventional safety hazards (and controls) and those which could contribute to a process safety event. This difference can be demonstrated using the task example of a pipefitter tightening a flange on a pipe. The conventional safety controls associated with the task include the following:

- Having correct isolations in place while performing the work to ensure the worker is protected from sources of energy and hazardous chemicals;
- Understanding and wearing the correct personal protective equipment;
- Having the correct tools for the task;
- Staying out of the line of fire while tightening the bolts using a wrench;
- Maintaining a tidy work area and being aware of the surroundings for slips and trips;

— Be properly tied off to anchor points if working from heights;
— Staying hydrated if working in hot or humid climates.

The process safety controls associated with the task include the following:

— Using the correct gasket material and shape;
— Using the correct bolts and torquing them correctly;
— Correct alignment of the pipes;
— Pressure testing the piping once completed.

In this example, failure of these process safety controls could contribute to a failure of the asset, which could lead to a more significant incident resulting in a catastrophic event or fatalities.

A system of hazard identification needs to be undertaken to identify the high level process safety risks during the engineering design of a uranium mine and processing facility, as well as during development of a comprehensive maintenance programme. This may include a hazard and operability ('HAZOP') studies, layers of protection analysis, historical incident reviews and brainstorming.

For new uranium production operations, the engineering design of the mine and processing facility is expected to consider and incorporate process safety in the early stages of hazard identification and risk analysis. Human factors need to be identified, evaluated and treated as part of the identification and management of process safety hazards. A plant layout study is beneficial to identify potential process safety events in relation to temporary or permanently occupied buildings. A safe and effective maintenance programme needs to be considered during the original design phase for the facility. This includes such aspects as location of process equipment, piping layout, valve placement and rigging points for equipment lifting.

Hazards that may be considered as process safety risks include:

— Explosions involving process materials or assets;
— Fires involving process materials or assets;
— Loss of containment of toxic, corrosive, reactive, asphyxiant, or hot materials, in bulk;
— Engulfment or physical impact from failure of bulk storage of process material(s) and process tanks;
— Structural failures of infrastructure;
— Exposure of the public.

The following systems will assist in managing process safety related risks:

— Pre-start safety reviews;
— Current and controlled operating procedures;
— Process safety critical controls;
— Process safety asset integrity;
— Quality design, installation and fabrication;
— MOC processes;
— Business resilience and recovery preparedness.

Several international reference documents have been developed to manage process safety related hazards, such as the following:

- — Australia's Standard for Control of Major Hazard Facilities [3];
- — China's Regulations on the Safe Management of Hazardous Chemicals [4];
- — European Commission Seveso Directive on Control of Major-accidents Involving Dangerous Substances [5];
- — Japan's High Pressure Gas Safety Act [6];
- — United States of America Occupational Safety and Health Administration Process Safety Management Regulations [7].

2.2. BOW TIE ANALYSIS

Bow tie analysis is a commonly used process safety risk identification and control method which highlights the threats, preventive barriers, hazards, recovery barriers and consequences, with a typical bow tie model shown in Fig. 1. Bow tie diagrams visually indicate the controls to prevent a safety related incident (e.g. primary loss of containment) on the left, and the controls to minimize the effect of the consequence on the right, after the event has occurred.

Bow tie analysis involves a multidisciplinary group from the engineering, operational and maintenance teams of a uranium production facility. The development of a bow tie diagram is associated with the documenting of a corresponding control activity sheet, where the risks, controls, responsible personnel and activity frequencies are determined, as shown in Table 1. Where applicable, the control activities can be integrated within a CMMS.

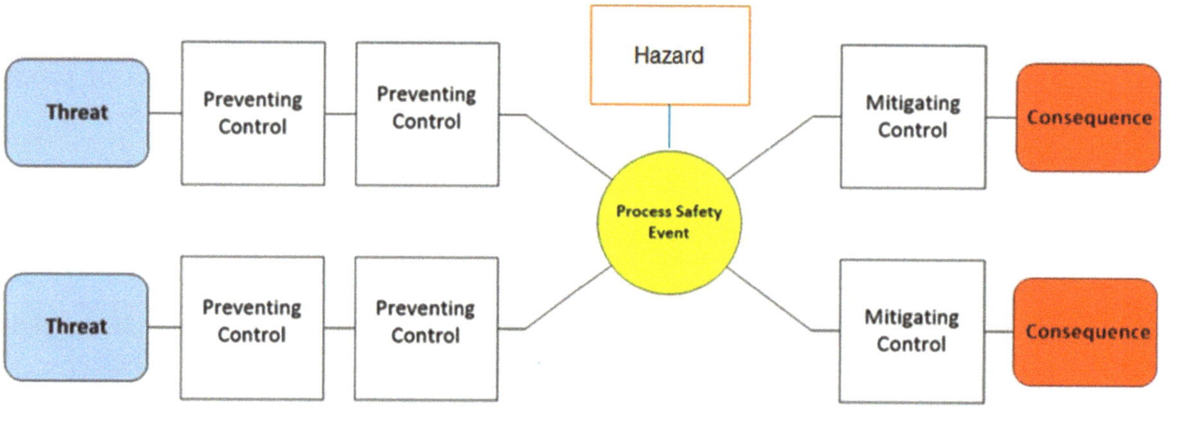

FIG. 1. Hazards and controls are represented diagrammatically on bow tie diagrams.

The main features of bow tie diagrams include:

- — Threats: Possible causes that potentially could result in a safety related incident (e.g. release a hazard) and produce a process safety event;
- — Preventing controls: Barriers that prevent threats (e.g. releasing a hazard) from occurring;
- — Hazard: Potential to cause harm, including injury and illness, damage to property, products or the environment, production losses or increased liabilities;
- — Mitigating controls: Barriers that act as recovery measures, which limit the impact of a consequence arising from the process safety event;
- — Consequences: Events that result from the release of a hazard.

Uranium processing facilities typically require sulphuric acid, ammonia and kerosene for extraction and purification of uranium. A typical bow tie analysis for ammonia and sulphuric acid is presented in Appendix I. Additional process hazards within a uranium production facility will be unique to the mining and processing activities, and may include hazards such as hazardous or flammable processing liquors, diesel fuel and uranium processing tailings.

The control activities are often monitored on a monthly or quarterly frequency. An example of a control activity sheet for an ammonia storage and distribution system (common to uranium processing facilities) is presented in Table 1.

Table 1 shows only the preventive controls, however a control activity sheet will document both preventive and mitigatory controls. Data to accompany each barrier (i.e. preventive and mitigatory controls) can include the following:

- Threat;
- Owner;
- Objective;
- Activities to achieve control objective;
- Who the activity is performed by;
- The method of verification.

TABLE 1. PREVENTATIVE CONTROL ACTIVITIES OF A TYPICAL AMMONIA BOW TIE ANALYSIS

Threat	**Preventative control**			**Frequency**	**Control owner**
	Critical control	**Hierarchy of control**	**Activity**		
Loss of vessel mechanical integrity due to corrosion or cracking	Vessel mechanical integrity	Administrative	Vessel inspection as per pressure vessel record book including thickness testing, weld/crack testing, maintenance plan reference numbers	Annually	Maintenance Engineering Lead
			Pressure test to 1.3 times maximum permissible working pressure. Maintenance plan reference numbers	Annually	Maintenance Engineering Lead
Loss of vessel mechanical integrity due to over pressurizing	Pressure relief valve (PRV) vessel	Administrative	Additional information: PRV specification	Annually	Maintenance Engineering Lead

TABLE 1. PREVENTATIVE CONTROL ACTIVITIES OF A TYPICAL AMMONIA BOW TIE ANALYSIS (cont.)

Threat	Preventative control			Frequency	Control owner
	Critical control	**Hierarchy of control**	**Activity**		
Loss of vaporizer and pipework mechanical integrity due to over pressurizing	PRV vaporizer and pipes	Engineering	Test and certification of vaporizer PRV. Pressure safety valve (PSV) has a unique identification number vaporizer set at 17.24 bar	Annually	Maintenance Engineering Lead
			Test and certification of pipework PRV's. PSV has a unique identification number. Applicable pressure relief valves, all set at 9.8 bar	Annually	Maintenance Engineering Lead
Loss of vessel mechanical integrity due to overfilling, over pressurizing, overheating, cooling the vaporizer	Vaporizer pressure, level and temperature alarm and interlock	Engineering	Level, pressure, temperature calibration and interlock test completed to include confirmation of high/low alarm at the central process control, confirmation of shut down and interlock functionality and maintenance plan reference numbers	Annually	Maintenance Engineering Lead

Each processing facility will have its own unique set of hazards, threats, barriers and consequences. As such, the examples provided are only indicative of what may be required for a uranium production facility. Other high risk processes that could result in multiple fatalities or widespread damage will be unique to the operation and need to be captured in separate bow tie analyses.

2.2.1. Fire risks and controls in solvent extraction circuits

Some uranium processing facilities use highly flammable organic solvents within the solvent extraction (SX) plant, which introduces a high fire risk. Examples of fires in uranium processing facilities include:

- 1978: Rössing Mine Namibia, Fire in solvent extraction delaying production by 1 year [8];
- 1999: Western Mining Corporation, Olympic Dam, AUD $10 000 000 damage, likely caused by static discharge of polyethylene pipe within the SX plant [9];
- 2001: Western Mining Corporation, Olympic Dam, AUD >$250 000 000 replacement cost, also likely caused by static discharge of polyethylene pipe within the SX plant [9, 10].

Historically, fires in solvent extraction circuits are shown to have a significant financial impact to the operation due to extensive repairs and lost productivity. In addition, fires in solvent extraction circuits present a significant risk to people and the environment, which need to be controlled. There are a range of preventive controls that can be implemented to minimize the risk of fire starting within a solvent extraction plant, as well as mitigatory controls to minimize the consequences if a fire were to start.

Preventative controls to minimize the risk of starting a fire include:

- Lightning protection at each corner and over the SX facility, with connection to an earth grid.
- Anti-static precautions, which include:
  - Use of earth straps between piping sections and correct earthing to ground, particularly for fibreglass, resin or high density polyethylene;
  - Earthing of portable pumps;
  - Intrinsically safe electrical equipment, including for motors and lighting;
  - Pneumatically or hydraulic driven equipment to minimize the risk of sparking or fire;
  - Synthetic clothing can be a fire hazard and is expected to be replaced with cotton or denim.
- The SX plant to be located away from vegetated land or other areas of the processing facility where fire could travel over to the SX plant.
- Grated flooring to prevent accumulation of solvent spillage.
- Expedient removal of organic spillage in bunds.
- Sealed spark proof motors.
- Fenced and secure areas to minimize access by personnel.
- Permits for hot work or use of certain tools within the SX plant.
- Signage and rules around smoking and risk of flames.
- Inductions and training for all personnel entering the SX plant.
- Well maintained equipment which is regularly inspected to minimize risk of fire hazard.

In the case of a fire starting, the following mitigatory controls would reduce the consequence of the fire:

- Well maintained fire detection system, which may include flame sensors that identify a live spark or fire and will trigger an automatic fire extinguisher system;
- Additional firefighting hydrants and equipment to allow fast response;
- Automatic fire suppressant foam deluge into SX mixers, settlers and process tanks that may contain solvents;
- Carbon dioxide extinguishers within buildings;
- Water curtains to protect control rooms or section areas of the plant;
- Fire walls between SX mixers, settlers and tanks including fireproof automatic closing doors to minimize risk of fire spreading to adjacent tanks;
- Fan intakes to switchgear rooms or motor control centres directed away from the facility;
- An emergency dump tank located at a distance away from the SX plant, to allow individual tanks to be quickly dumped away from the fire hazard;
- Equipped and trained emergency services teams.

Fire risks are a significant hazard and could be included as a major process safety risk. An SX fire hazard bow tie for a uranium processing facility has been included in Appendix I.

2.3. MANAGING NON-CONFORMANCE AND AUTHORITY TO OPERATE

An approval process is needed for continued operation of an asset outside its planned maintenance schedule or with a known defect. This is commonly referred to as an authority to operate (ATO). All critical equipment which is operated with non-conformances (i.e. defects) or operating while deviating from scheduled internal inspections or licence conditions, need to be recorded and managed under an ATO process. This process may be guided by national and local regulations or through the facility's permit to operate conditions. It is good practice to develop a register (or database) of ATO documents to more easily track open ATOs.

The ATO process relies on a system of defect identification and notification. Defects on critical equipment are most commonly identified during periodic inspections or by production operators (also known as plant operators) during normal operation (i.e. a leak detected in a vessel). Identified defects need to be escalated to the maintenance engineering team, through a formal notification system, to initiate the ATO process. Until this notification has occurred, the responsibility for safe operation of the asset remains with the department responsible for the operation of the asset.

All risks associated with the operation of a non-conforming asset need to be identified, and controls implemented to mitigate the risk (see Section 2.5). Only once the risks are mitigated to as low as reasonably achievable, with controls in place to monitor, maintain or reduce the risk level, can the ATO be approved by both technical experts and operations management. In some operations, a process to escalate the approval of certain high risk scenarios to senior management (e.g. general manager level) may be required, as determined by the site specific conditions. The ATO approval will only be valid for a set period, as informed by the risk assessment.

In some circumstances, a non-compliance with a standard may be found however the risk associated with the non-compliance does not necessitate corrective action. This is a common scenario within ageing uranium processing facilities and may include things like non-compliant design, construction or repairs, where standards have changed. In these situations, approval is gained from a technical expert and documented within a risk assessment, for the continued operation of the non-compliant asset.

2.4. CHANGE EVALUATION AND MANAGEMENT

IAEA Nuclear Energy Series NG-T-1.1, Managing Organizational Change in Nuclear Organizations [11] states:

> "introducing any change in a nuclear facility has the potential to have an impact on safety; therefore, it is essential to have processes that support the effective management of change. These processes enable the facility to identify the risks associated with a change and to manage the risks properly while realizing the benefits of the intended change."

Typically, change can be organized into two main categories:

(1) Organizational change that involves significant organization wide change, such as a reorganization or adding a new product or service;

(2) Smaller changes such as adding a new person, plant modifications, minor alterations to a programme or adopting a new procedure.

Henceforth, this publication will consider further only the change management process for smaller modifications, as it relates specifically to asset management in ageing uranium mines and processing facilities. Detailed guidance on identifying, planning and implementing organizational change in nuclear facilities, can be found in Ref. [11].

Changes to the assets, asset management and asset management systems in uranium mines and processing facilities can be triggered by external and internal factors, such as new legal requirements, plant modifications, or process alterations. For example, revised or new industry safety standards may necessitate changes to uranium mining and/or processing facilities and their operations. New technologies or innovations may be implemented to improve uranium processing. Or any myriad of modifications relating to plant configuration such as alterations (or replacement[1]) of structures, systems or components, process software, operational limits and conditions, operating procedures, and plant drawings [12].

The risks associated with any planned change, permanent or temporary, that can have in impact on achieving asset management objectives, need to be assessed before implementing the change. This is termed 'management of change'[2] (MOC) and Ref. [14] states:

> "risks associated with changes should be evaluated and managed to ensure that the integrity of the asset management system is maintained and risks arising from changes remain at an acceptable level. Changes can be in: asset portfolio; asset management practices/activities; the asset management system processes; the structure of the organization. An organization should establish processes and programmes for the management of both temporary and permanent change."

A thorough MOC process assesses the risk of the change not only on the asset management system objectives, but on other areas of the organization, including financial, safety, productivity, environmental, community impact, etc., to determine if the change will increase or introduce new risks in these areas. A risk assessment process is described further in Section 2.5.

An effective MOC process is essential to the integrity of the asset management system, irrespective of whether the change is permanent or temporary, gradual or sudden, planned or unintended. Good industry practice is for an organization to implement an MOC procedure and form that allow the details of the change to be recorded, assessed and approved, and with an action tracking (and subsequent close-out) process. Key information to record is as follows:

- Change title, date, reference number;
- Type of change (emergency, permanent, temporary, trial, contingency);
- Emergency change authorization (typically senior management);

[1] This does not include replacement of a component by an equivalent component in a recognized maintenance activity, for example one that is identical to the original components or has the same design requirements.

[2] It is noted that there is transition from 'management of change' to 'control of change' in the 2024 release of the International Standards ISO 55001:2024 Asset Management – Management system [13], which is not yet reflected in Ref [14].

— Location of the change (area or equipment number);
— Change purpose;
— Change description (drawings to be added if required);
— Expected outcomes of the change;
— Measures used to monitor the change;
— Lists all documentation associated with the change (may also be attached);
— Actions required if change implemented;
— Approval to proceed with change;
— Close out of MOC once all actions completed;
— Quality review of the effectiveness of the change process.

The intent of an MOC process is for people with the relevant experience, knowledge, qualifications and accountability to inform the process to deliver quality outcomes.

A key consideration in the MOC process is to identify relevant stakeholders (i.e. those who need to be consulted to help identify potential hazards, assess risks and identify controls). The organization may consider including a prompt to identify key stakeholders in MOC documents. An example list of prompts for an MOC form that may be useful for a uranium mine or processing facility is included in Appendix II.

The relevant stakeholders need to be informed about changes prior to being implemented. If the engagement with relevant stakeholders concludes that the change introduces unacceptable risk — for example, to the safety, environmental or financial performance of the mine or processing facility — then the change cannot be implemented.

While anyone in an organization may initiate a change, the system of MOC requires accountability of persons within the organization to coordinate and lead the process. An example of good industry practice is to assign an MOC Owner, for each proposed change. Provided below is an example of the purpose and accountabilities of the role of MOC Owner. The selection of individuals to perform the role of MOC Owner will be unique to the change itself. The organization needs to provide training on the MOC process to the MOC Owner. Generally, within the organization there are personnel trained in each department to provide support for MOC.

Typically, the MOC Owner is the person responsible for coordinating all aspects of the change, concluding the change and ensuring requirements are met. This person facilitates the process to ensure all risks associated with the change are identified. This person also ensures that each component of the MOC is adequately addressed, documented, that approvals and authorizations are sought, and tracks actionable items through to completion stage.

The MOC Owner decides whether technical reviewers are required and if so, assigns these reviewers. Ideally changes are technically reviewed by personnel from multiple disciplines (e.g. engineering, legal, procurement, occupational health, safety and environment).

Importantly, the MOC process includes senior management oversight and a formal system of approval.

As a result of change management, records, drawings, specifications, procedures or other documentation, may require revision and knowledge management, as outlined in Section 5.5, of this publication.

2.5. RISK ASSESSMENT

The development of an asset management programme (see Section 4) requires an understanding of the risks associated with the failure of each piece of equipment. Understanding how the failure of equipment affects the business operations (e.g. plant efficiency, production costs, safety performance, etc.) will inform if more or less maintenance is to be applied to that equipment. This section outlines a general process of preparing for and performing a risk assessment, which can be used to perform an asset criticality assessment (ACA), as well as for assessment of risk in many other areas of the operation.

An organization responsible for uranium mining and processing is expected to establish a risk assessment process that can be used in all aspects of the business, including asset management, worker safety, financial, etc. It is important that the process is well defined and embedded into the operations, such that all risks are assessed equally using the same descriptors and assessment parameters. Assessment of risk needs to involve people with the relevant experience, knowledge, qualifications and accountability to inform the process to deliver quality outcomes.

A measure of risk can be obtained through a process of assessment of the consequence (or impact) of an undesired scenario and the likelihood of that undesired scenario occurring. The combination of consequence and likelihood factors are then used to determine the level of risk. The following sections will outline how to perform a risk assessment. Generally, there are three types of risk assessments:

- Qualitative risk assessments: More subjective and rely on the knowledge and experience of personnel performing the risk assessment to determine the likelihood and consequence.
- Semi-quantitative risk assessments: Primarily a qualitative risk assessment which is supported by available data to help improve the accuracy of the assessment.
- Quantitative risk assessments: Uses numerical estimations and verifiable data to specify the impact (i.e. production costs, down time, etc.) and likelihood (i.e. incident frequency rates) in absolute values. Conducting these risk assessments are more time consuming and require a detailed understanding of the consequences with data available.

Qualitative risk assessments are more commonly used as they provide an assessment of the risk with sufficient accuracy of the descriptions relative to the scenario. A quantitative assessment is typically used for high risk scenarios where a more thorough assessment is needed.

2.5.1. Consequences of failure

An undesired scenario may have a range of consequences from financial, safety, productivity, environmental, community impact, etc. Similarly, the impact or consequences of equipment failure needs to be assessed against these consequence criteria.

The severity of the consequences of failure can be determined through comparing the consequence against a range of consequence descriptors. The most common categories are occupational health, safety, production volumes, the environment and the community. Other categories may include descriptors for capital expenditure, schedule, revenue, operating cost, compliance, business reputation or any other aspects that needs to be assessed. Each of these categories will have a set of descriptors that describe the severity of the consequences. The consequence categories and descriptors will vary significantly between industries and organizations.

The consequences of equipment failure are specific to the operating environment, surrounding area or system the asset is operating in. For example, the consequences of failure of a large water tank at a remote location would be far less than the same failure adjacent to a heavily trafficked area or where secondary damage can occur. Furthermore, the consequences of failure of a piece of equipment may have a high production consequence if it stops production. However, this same piece of equipment could have a low production consequence if in a duty/standby arrangement, meaning it can be changed out or repaired with minimal production down time.

Organizations responsible for uranium mining and processing need to establish consequence categories and descriptors which are suitable for conducting risk assessments for such facilities and activities. The organization needs to utilize consequence descriptors within the framework of an established risk assessment process to ensure that risk assessments are completed consistently.

An example of a typical consequence descriptor table for a uranium mining and processing facility is shown in Table 2.

TABLE 2. EXAMPLE OF CONSEQUENCE DESCRIPTORS FOR A URANIUM MINING AND PROCESSING FACILITY

Consequence	Production volumes	Total financial/ revenue	Safety	Occupational health	Environment
Minor	Impact on production <0.5%	<2%	Near miss	Reversible health effects of little concern	Immediate reversible impact
Medium	Impact on production 0.5–2%	2–5%	First aid	Reversible health effects of concern	Short term reversible impacts
Serious	Impact on production is 2–5%	5–10%	Medically treated injury	Severe reversible health effects of concern	Medium term recovery impacts
Major	Impact on production is 5–10%	10–20%	Lost time injury	Single fatality or irreversible health effects	Long term recovery leaving residual damage
Catastrophic	Impact on production is >10%	>20%	Fatality	Multiple fatalities or serious disabling illnesses	Widespread impact, long term recovery, major residual damage

During the risk assessment process, site based operations personnel familiar with the relevant aspects of the uranium mining and processing facilities and activities need to be involved. The personnel involved need to be well positioned to understand the consequences of the failure scenarios being assessed, providing valuable input into the selection of the appropriate consequence categories and consequence levels.

During the risk assessment process, the maximum reasonable outcome (MRO) needs to be selected, which is the maximum consequence scenario that could reasonably be expected to occur.

2.5.2. Likelihood of failure

The likelihood of an undesired scenario, such as equipment failure, describes how frequently the event or scenario is expected to occur. Similar to consequence descriptors, likelihood descriptors vary significantly between industries. The organization needs to establish, document and utilize a standardized likelihood descriptor table such that all risk assessments are completed consistently.

The likelihood of equipment failure needs to be assessed for each relevant consequence scenario identified. Historic data (i.e. age, performance, reliability, design parameters) of the asset or similar assets can assist with this assessment of likelihood of failure. Institutional knowledge is particularly helpful in this regard, where documented or first-hand information about the asset can be used to establish the likelihood of a failure. An experienced engineer can assist with likelihood assessments and provide further guidance on the remaining life of the asset. Information from external sources can also assist to provide an estimate of the likelihood of failure (e.g. industry publications, government resources on incidents).

An example of a likelihood descriptor table for a uranium mining and processing facility is shown in Table 3.

TABLE 3. EXAMPLE OF POSSIBLE LIKELIHOOD DESCRIPTORS FOR A URANIUM MINING AND PROCESSING FACILITY

Descriptor	Broad description	Timing descriptor
Almost Certain	Happens often. Has previously occurred monthly in this organization	More than 1 event per month
Likely	Could easily happen. Has occurred in the last few years in this organization or has occurred recently in other similar organizations	More than 1 event per year
Possible	Could happen. Has occurred at least once in the history of this organization	1 event per 1–10 years
Unlikely	Hasn't happened but could. Has never occurred in this organization but has occurred infrequently in other similar organizations	1 event per 10–100 years (i.e. within a single mine life)
Rare	Conceivable, but only in extreme circumstances. Is possible but has not occurred to date in any similar organization	Less than 1 event per 100 years

2.5.3. Risk matrix

A risk matrix is a tool that links the consequence and likelihood of the undesired scenario to obtain the overall level of risk. Descriptors to rate the level of risk need to be established by the organization. Commonly used risk rating descriptors are Low, Moderate, High, Critical, or else numerical descriptors (e.g. 1–25) are used. The risk ratings assist the owner/operator to determine if the level of risk of the undesired scenario is acceptable to the organization, or if additional controls need to be implemented to mitigate the risk to an acceptable level. Similar to the consequence and likelihood descriptor tables, each organization will have their own risk descriptors and risk matrix which need to be applied to all risk assessments to ensure consistency across its operations. An example of a risk matrix is shown in Fig. 2.

		Consequence of Failure				
	Descriptor	Minor	Medium	Serious	Major	Catastrophic
Likelihood of Failure	Almost	Moderate	High	Critical	Critical	Critical
	Likely	Moderate	High	High	Critical	Critical
	Possible	Low	Moderate	High	Critical	Critical
	Unlikely	Low	Low	Moderate	High	Critical
	Rare	Low	Low	Moderate	High	High

FIG. 2. Example of risk matrix.

An example of risk descriptors is shown in Table 4. These are again specific to each organization. For the purpose of performing an ACA in support of the development of an asset management programme, an additional column has been included to show how the criticality assessment level can correspond to the risk rating.

TABLE 4. EXAMPLE OF RISK DESCRIPTORS

Risk rating	**Asset criticality assessment**	**Description**
Critical	Level 4	Risks that significantly exceed the risk acceptance threshold and need urgent and immediate attention
High	Level 3	Risks that exceed the risk acceptance threshold and require proactive management
Moderate	Level 2	Risks that lie on the risk acceptance threshold and require active monitoring
Low	Level 1	Risks that are below the risk acceptance threshold and do not require active management

Once the likelihood and consequence descriptors are determined for each undesired scenario, they can be plotted on the risk matrix to determine the overall risk for that scenario. The risk matrix example shown in Fig. 3 illustrates that if the likelihood of failure is 'possible' and consequence of the failure is 'serious' then the level of risk will fall into the category of 'high'.

		Consequence of Failure				
	Descriptor	Minor	Medium	Serious	Major	Catastrophic
Likelihood of Failure	Almost	Moderate	High	Critical	Critical	Critical
	Likely	Moderate	High	High	Critical	Critical
	Possible	Low	Moderate	High	Critical	Critical
	Unlikely	Low	Low	Moderate	High	Critical
	Rare	Low	Low	Moderate	High	High

FIG. 3. Example of risk matrix showing level of risk.

When using the risk assessment process to assess hazards within the facility or tasks to be performed, all risks are to be reduced to as low as reasonably achievable. High risks are generally not acceptable and additional controls are to be implemented to mitigate the risk. The results of assessment of risk for each scenario, including the identified control measures, need to be documented, for example in a risk register.

Senior management of organizations need to be involved in the assessment of risks to determine which risk levels require additional controls or contingency plans to lower the risk. This is to ensure that senior management are aware of the high risk scenarios or task to be performed and can intervene if necessary. Additionally, a process of senior management authorization, in the form of a signed approval, may need to be considered if the level of risk cannot be further reduced.

Section 4.2 further describes how the risk assessment process outlined above can be used to determine strategies for life cycle asset management in uranium mining and processing facilities, through an ACA.

For further information on risk analysis, the following publications have been developed as guidelines for conducting risk assessments:

- The British Standards Institution, standard BS EN 31010:2019 for Risk Management: Risk Assessment Techniques [15];
- The United Kingdom Health and Safety Executive, discussion document on Risk Criteria for Land-use Planning in the Vicinity of Major Industrial Hazards [16] and Five Steps to Risk Assessment [17];
- The American Institute of Chemical Engineers, Guidelines for Chemical Process Quantitative Risk Analysis [18];
- The Committee for the Prevention of Disasters, Guidelines for Quantitative Risk Assessment [19].

2.6. RADIATION PROTECTION FOR MAINTENANCE WORK

Maintenance personnel working in uranium mines and processing facilities are often working in proximity to radiation sources such as the uranium ore, process slurries and solutions, uranium ore concentrate and uranium tailings. A radiation protection programme needs to be established in compliance with the State's legislative requirements and provisions for occupational exposure control, monitoring, assessment and recording of employees of the facility, as well as for contractors performing planned maintenance work [20, 21]. In the case that maintenance worker are not employees of the management of the facility (i.e. maintenance workers employed by a contractor providing services) special arrangements need to be made for radiation protection under a contractual agreement [20, 22].

In uranium mines and processing facilities, radiation protection is part of the overall management system for occupational health and safety required for the protection of workers from ionizing radiation. The management systems for these facilities is expected to be designed to implement radiation protection and safety measures for personnel performing asset management activities and apply the optimization principle [20, 21]. The organization is expected to establish procedural and technical arrangements for the designation of controlled areas and supervised areas, for local rules and for monitoring of the workplace, in a radiation protection programme for occupational exposure [20].

When mining and processing equipment are being prepared for maintenance work, standard procedures include isolation from both non-ionizing energy (e.g. electrical or mechanical) and ionizing energy sources of the equipment to be maintained. This includes different exposure pathways including external exposure to gamma radiation, inhalation of radon and radon progeny and inhalation of long lived radionuclide dust. For example, it is standard practice to ventilate, flush and clean equipment for control of, and to optimize, the exposure [21].

Where maintenance workers might be exposed to sources of gamma radiation in the mine or processing facility, a maintenance work plan needs to be developed and implemented to ensure the work is completed in a safe and efficient manner and the maintenance worker does not exceed radiation dose constraints established by the operator and dose limits according to the national legislation. When developing the maintenance work plan, radiation protection measures to reduce workers' dose, the hierarchy of controls approach to instituting controls [21] and the radiation monitoring requirements, are to be considered.

The general radiation protection measures to be considered in a maintenance work plan will include:

— Requirements for maintenance workers to be provided with radiation training and fitted with appropriate personal protective equipment.
— Prior to maintenance workers arriving at the work face, the equipment is to be isolated, ventilated, flushed and cleaned.
— Good industry practice is to have all necessary parts (e.g. tools, spares, etc.) required for the maintenance task to be provided at the work face. This will improve the efficiency of the maintenance work, and ideally reduced the total time required to complete the task.
— Requirements for maintenance equipment and tools to be cleaned by the maintenance workers, and in some cases radiation clearance protocols to be completed with a radiation protection officer;
— Radiation monitoring to be completed before, during or after the maintenance work, by a qualified and trained radiation protection officer;
— Maintenance workers being provided with personal radiation dosimeters.

The selection of monitoring dosimeters for individuals or similar exposure groups is informed through radiation monitoring and surveys by the organization's appointed radiation protection officer. Passive devices (e.g. thermoluminescent dosimeters) are commonly issued to workers for personal radiation dose monitoring.

Reference [21] states:

> "The primary control method for the mining and processing of high grade ores is isolation of the material from the workforce. In practice, this means a strong commitment to radiation protection being an integral part of both the design and day to day operation of the facility. However, a lack of appropriate controls has the potential to result in situations where doses approaching or exceeding the occupational dose limits could occur in a relatively short period of time (e.g. days to weeks)."

If a maintenance worker is approaching an assigned limit on their exposure, in accordance with the facility's radiation protection programme, the maintenance worker will need to leave the work face and report to their supervisor. The supervisor, in consultation with a radiation protection officer, has responsibility to reassign that worker to an area of the facility that separates the worker from exposure to radiation sources. In order to complete the planned work, the supervisor needs to reassign a new maintenance worker to the maintenance task in progress, with additional control measures to ensure that their radiation dose remains below their allowable limit. In some high grade uranium processing facilities, the maintenance workers may only be able to stay in the area for short durations (e.g. one to two hours) before they have reached their daily radiation dose limit and need to leave the area. These high radiation areas are usually towards the back end of the processing plant where uranium is becoming more highly concentrated or where uranium progeny are concentrated such as tailings slurries. Scale inside tanks and pipes can also be highly radioactive and need to be surveyed and managed accordingly.

Performing maintenance activities in uranium calciners or roasters may require normal operational radiation control measures to be bypassed, for example isolation of electrical sources may shut down ventilation systems, or maintenance personnel may be required to enter 'restricted access' areas (e.g. sealed U_3O_8 product drumming rooms) to perform the maintenance. Alternate control measures for maintenance activities need to be defined in the radiation protection programme, to manage potential exposure to yellow cake dust and calcined concentrates. Additional measures may include considerations in planning, enhanced personal protective equipment (PPE), work execution and post-maintenance cleaning.

Access to equipment and infrastructure, in the area where the calciner or roaster is located, will likely require personal respiratory protection and an additional layer of protective clothing, to prevent inhalation of uranium dust and the surface contamination of work clothing, respectively. Examples of respiratory protection that may be used in this environment include powered air purifying respirators with the appropriate air filtering cartridges. Supplied air breathing apparatus or self-contained breathing apparatus may also be used for certain maintenance activities. These types of personal respiratory protection devices provide enhanced respiratory protection from uranium dusts when compared to passive half mask air purifying respirators. Workers will require training in the use of this type of personal respiratory protection. Disposable coveralls can be worn over top of regular work coveralls to enhance hygiene and prevent calcined uranium tracking and contamination of the worker and work areas outside of the calciner or roaster.

Entry and exit procedures are also to be considered as part of the radiation protection programme for maintenance work in a uranium calciner/roaster. An example of good practice steps for entry into the calciner/roaster area for maintenance works are provided below.

(1) The maintenance worker receives a radiation work permit, issued by the radiation protection officer, for the planned radiation exposure due to working in the calciner/roaster. The work permit describes the radiation control measures for the work. The maintenance worker may also be issued with a dust monitoring pump for radiation dose control purposes.
(2) If applicable, the maintenance worker submits a pre-entry uranium-in-urine bioassay sample. This will largely depend on the regulatory requirements and/or the requirements of the individual site, as some mines only conduct monitoring in the event of a suspected incident.
(3) The maintenance worker proceeds to a radiation clean work zone to put on disposable coveralls, rubber boots, respiratory protection equipment, hard hat and gloves. The dust monitoring pump is added with the pump assembly inside the disposable coveralls and the filter located near the breathing zone. Further, it is good practice to seal all interface points (e.g. between respiratory protection equipment and the hood of the disposable coveralls and gloves/boots and disposable coveralls) with tape to reduce the risk of calcined uranium ingress and exposure.
(4) The maintenance worker is now ready to commence the maintenance task.
(5) Following the completion of the maintenance task, the worker enters a decontamination area where the worker showers with all of their protective equipment still on. This removes any surface calcined uranium and prevents inhalation, ingestion or tracking of this material. It is good practice to locate the wash down area next to the calciner/roaster area but outside of the process where this equipment is located. A drain system is needed to collect the calcine loaded wash down water and this water is redirected back to the hydrometallurgical process.

(6) The worker can now proceed to an adjacent disrobing room where they can remove their gloves, disposable coveralls, rubber boots and lastly their respiratory protective equipment. A procedure may be developed to manage this used PPE where it can either be cleaned for reuse or disposed of.
(7) The worker returns the signed off work permit to the radiation department along with the dust monitoring pump.
(8) The worker then proceeds to the clean shower area and has a shower prior to submitting a post-work uranium-in-urine bioassay sample;
(9) The pre- and post-uranium urine samples are then analysed for uranium to determine the concentration of uranium in the urine before and after entry into the calciner/roaster work area. If ingestion or inhalation of uranium are detected based on pre- and post-uranium-in-urine monitoring, then the maintenance worker needs to be placed on an enhanced monitoring programme that may include subsequent urine analysis, whole body counting or simulated lung fluid analysis [21]. The regulatory body and/or the occupational health physicist (if applicable) can develop further protocols on control measures and exposure limits to be incorporated into the radiation protection programme.

IAEA Safety Standards Series No. GSR Part 3, Radiation Protection and Safety of Radiation Sources: International Basic Safety Standards [23] establishes requirements that apply to all facilities and activities that give rise to radiation risks, for the protection and safety of people and the environment. Recommendations on the development of occupational radiation protection programmes are provided in IAEA Safety Standards Series No. GSG-7, Occupational Radiation Protection [20], and further information specific to occupational radiation protection in the uranium mining and processing industry is provided in Ref. [21].

3. OVERVIEW OF ASSET LIFE CYCLE MANAGEMENT

An effective asset management strategy is an essential aspect of uranium mining and processing operations from economic, safety, environmental, sustainability and public acceptance perspectives. Effective strategies and programmes that focus on failure and loss elimination will ensure safe and sustainable production through the life cycle of the mine and processing facility. Asset management strategies need to be developed prior to commissioning and operating a uranium mine and processing facility. Development of this strategy begins once detailed engineering design for the uranium mine and processing facility is complete and the subsequent drawings for construction have been issued. At this point the owner/operator of the uranium mine and processing facility will have detailed knowledge of the type of assets (e.g. infrastructure and equipment) that will be utilized in these facilities.

Development of an asset management strategy involves the following four generalized steps:

(1) Implement a systematic approach to cataloguing all equipment and infrastructure into a database.
(2) Develop and utilize a standardized risk assessment tool to categorize and rank all assets, if a specific piece of equipment or infrastructure fails. This process will assist in identification and prioritization of asset management strategies for high risk equipment and infrastructure.
(3) Utilize a tool such as a failure modes and effects analysis (FMEA) to understand the failure modes of high risk equipment and infrastructure.

(4) Use the results of the FMEA to develop equipment and infrastructure specific asset management activities that will facilitate safe and sustainable operation of the uranium mine and processing facility.

An asset management strategy requires an integrated approach, that is supported by a CMMS for cataloguing all equipment and infrastructure, and for predictive maintenance and inspection based preventive maintenance programmes [24].

All assets will fail at some point. However, it may take a long time (i.e. months) between an initial sign of failure and the complete failure of the asset. An effective asset management programme will ensure defined assets are required to be repaired or replaced before complete failure.

Industry experience has shown that effective predictive maintenance strategies have the ability to increase the duration between maintenance repairs of assets, when compared to preventive maintenance strategies. Predictive maintenance strategies provide longer warning periods before the complete failure of an asset, making it easier to plan and schedule for repair or replacement. In addition, asset management strategies that involve predictive maintenance monitoring often have an increased ability to accurately define failure modes when compared to using only preventive maintenance practices. This is due to the extended run time between failure for predictive maintenance strategies. Technologies that may be employed in predictive maintenance include vibration analysis, oil analysis, thermography (e.g. bearing temperature) and acoustic analysis (e.g. bearing noise).

The Potential-Failure (P-F) curve shown in Fig. 4 illustrates the interval between a potential failure (i.e. when failure can be observed or detected for the first time) and the functional or complete failure of the asset. The curve provides examples of key indicators of failure along the curve which can be monitored, measured or observed.

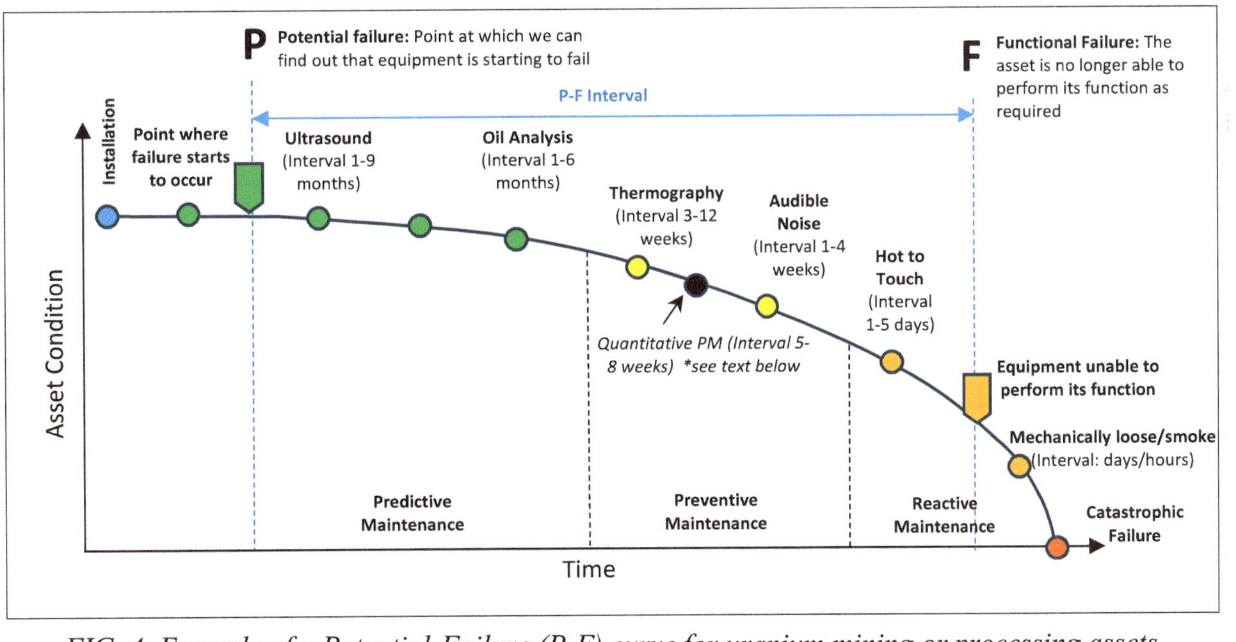

FIG. 4. Example of a Potential-Failure (P-F) curve for uranium mining or processing assets.

In general, preventive maintenance tasks have been shown to be less effective in extending the duration of operation before asset repair or replacement. For example, the black dot in Fig. 4 (not typically in a P-F curve) was added to illustrate that preventive maintenance tasks and equipment repairs often occur too early, which results in increased life cycle costs and increased downtime. In this example, the run time of the equipment may have extended for an additional 5–8 weeks before having to repair or replace the asset. Extending the run time facilitates increased productivity, as the equipment is monitored while it is operating, lowers life cycle costs and may also provide additional information on failure modes.

From an economic perspective, once implemented, predictive maintenance monitoring of assets is shown to reduce the maintenance costs of an asset by up to 75%, as the quantity of labour intensive preventive maintenance work is significantly reduced. Furthermore, preventive maintenance work may introduce maintenance induced failures that may not have otherwise occurred. There is an initial higher cost with developing predictive maintenance strategies as specific monitoring instrumentation needs to be installed on relevant assets. It also requires trained staff to monitor and interpret the data collected from the asset. To determine the best maintenance strategy for each asset, experienced and qualified asset management personnel need to complete a cost benefit analysis comparing predictive maintenance, preventive maintenance and run to failure (see Section 4.3). In general, the cost to maintain an asset ought to be lower than the cost (or impact) of the failure. The hierarchy of an asset management strategy, from an economic and technical perspective, can be prioritized and evaluated in the following order:

(1) Process monitoring (including asset operation inspections);
(2) Predictive maintenance strategies;
(3) Metering/time based directed tasks (preventive maintenance).

In some cases, a preventive maintenance strategy is more economic and technically compliant. An example of this may be a slow rotating asset where it makes sense to take the asset offline at a prescribed frequency, complete an inspection and replace the bearings. The used bearings could be inspected for wear and the preventive maintenance frequency adjusted (e.g. extended) based on the condition of the bearings. This strategy may be more economic than installing and maintaining monitoring equipment for predictive maintenance.

A well developed and implemented asset management strategy leads to an increase in mine and processing facility availability for uranium production activities, which can be measured as an improvement in overall equipment effectiveness. It also results in an overall reduction in life cycle operating costs. In addition, a well executed asset management strategy has the potential to increase the safety performance of the mine and processing facility as there are fewer unplanned failures and abrupt processing stoppages, which may also result in increased radiation exposure and increased conventional safety risk. Finally, an effective asset management strategy also results in reduced unplanned or emergency maintenance work that can lead to increased safety risks and operating costs (e.g. reduced focus on safety in order to get production back on-line and increased costs with expedited parts and overtime payment).

Overall, an effective asset management strategy will result in reduced downtime, lower labour costs, less rework and overall lower life cycle costs of an asset and enhanced safety and environmental performance of a uranium mine or processing facility through the lifetime of the facility. Sections 4–7 describe in more detail the specific aspects, steps and processes involved in developing an effective asset management strategy.

3.1. ASSET MANAGEMENT FRAMEWORK

Uranium mines and processing plants are complex facilities that require organizations conducting uranium production activities to have a management system in place that integrates all relevant elements, including safety, health, radiation protection, radiation safety, environmental, security, quality, human-and-organizational-factors, societal and economic aspects [2]. As part of the integrated management system, an asset management system for ageing facilities, needs to be put into effect by the operating organization, as an integral part of safe and efficient operation of a uranium mine and processing facility [24].

Requirements for the management system are established in IAEA Safety Standards Series No. GSR Part 2, Leadership and Management for Safety [2], and recommendations are provided in IAEA Safety Standards Series No. GS-G-3.1, Application of the Management System for Facilities and Activities [25].

According to the International Organization for Standardization (ISO), the international standard for asset management can be summarized as the set of coordinated activities that an organization uses to realize value from assets in the delivery of its outcomes and objectives. Realization of value requires the balance of costs, risks and performance of each asset, over the life of the facility [13, 26].

Reference [26] provides guidance on asset management and asset management systems, with application to a broad range of assets, and organizations, including the uranium production industry. The standard is supported by Ref. [13] which specifies the requirements for an asset management system, and Ref. [14].

An asset management system is supported by an asset management framework for maintenance of assets, such as those found in uranium mines and processing facilities. The Global Forum on Maintenance and Asset Management has been established with the aim of sharing collaboratively on advancements, knowledge and standards in maintenance and asset management. The Forums 'Maintenance Framework' provides a good industry example of asset management framework, that ultimately supports the objective of a decision making process that aligns asset maintenance delivery with the corporate objectives. The framework is intended to lead to value creation of assets and ensures the health and safety of workers and the public are maintained [27]. The management framework (adapted) can be summarized into seven elements:

(1) Organization requirements;
(2) Maintenance strategy;
(3) Maintenance management;
(4) Maintenance execution;
(5) Continuous improvement;
(6) Allocation of resources;
(7) Deliverables.

Additional detail on each of the elements of an asset management framework is provided below, in the context of an organization mining and processing uranium.

3.1.1. Organization requirements

The assets of a uranium mine and processing facility need to be matched with the organization's requirements. Therefore, the operational plans to achieve the organization's objectives need to be understood and integrated into the asset management strategy. For example, the asset management strategy would consider the expected lifespan of the facility which is typically estimated as part of the economic model for developing and operating a uranium mine and processing facility. Ore grade, environmental conditions, metallurgical processes, planned production rates, and many other aspects are considered in the design of a facility's assets. Effective asset management will consider how to efficiently maintain assets and provide best value over the asset life cycle, while meeting the organization's requirements.

3.1.2. Maintenance strategy

Specific maintenance strategies are designed for each asset and will vary depending on the criticality of that asset to the mine or processing facility, the design of the uranium extraction process, and the negative impacts of asset failure. If the criticality of the specific asset is determined to be minor in nature, then there is an opportunity to apply a common, and generic, 'asset type' strategy. An example of this would be one common maintenance strategy for all the lighting systems at the facility where these components would be run to failure (e.g. replaced light bulb when burned out). As the criticality of the asset increases, or the impact of failure becomes more significant, then the asset criticality is analysed, and an FMEA completed to help determine an appropriate maintenance strategy for that asset. Detailed information on assessing asset criticality and failure modes analysis in developing an asset management programme is provided in Section 4.

3.1.3. Maintenance management

The Global Forum on Maintenance and Asset Management describes maintenance management as the "process to assure that assets continue to do what the organization business requires both in the present and future operating contexts" [27]. It recognizes that an organization's processes and practices for leadership management, planning, scheduling, preventive maintenance, condition monitoring, maintenance tasks execution, record keeping, data management, failure analysis, and spare parts management are expected to demonstrably deliver on the businesses needs and outcomes.

Once a uranium mine and processing facility is constructed, commissioned and operational, the routine management of critical assets is necessary in order to monitor their condition and to identify issues of concern in a timely manner. Asset condition monitoring may include inspections, predictive maintenance techniques, and preventive maintenance routines. Detailed information in maintenance management tactics development is provided in Section 4.3.

3.1.4. Maintenance execution

Maintenance execution encompasses the process of completing work within the maintenance system. Section 5 provides detailed information on maintenance execution, work management process and performance management. Most organizations utilize a CMMS to facilitate the maintenance workflow. Within these systems are the processes that allow work to be identified and planned, spare parts to be purchased and prepared, the work to be scheduled, directions for tradespersons to execute the work, and financial costs to be captured and allocated. The CMMS may also be used to gather and store information once a maintenance task has been completed. Information may include feedback from the trades and actual time required to complete the task, reports on condition of components evaluated during the maintenance task (e.g. bearings, etc.) and any aspects that were missed or inaccurate in the original plan, scheduling and execution. This data can then be used to modify and continually improve the asset management programme including maintenance frequency, planning, scheduling, tooling and work execution.

3.1.5. Continuous improvement

Each of the elements of the framework, as described above, can be improved and optimized in order to realize best value, this is commonly referred to as 'continuous improvement' [13] and follows the principle of the 'Plan – Do – Check – Act' cycle [13, 28]. The asset management strategies are to be reviewed periodically to ensure the original maintenance tactics employed are still the best tactics to use, and take into consideration the current needs of the organization, changes to local environment, social factors, and technology improvements. Defect elimination, which refers to tracking and studying equipment failures and then improving the maintenance strategy to avoid those failures in the future, is an example of one of the mechanisms that can be use used to continuously improve asset management strategies. Section 5.3 explores performance management and continuous improvement in more detail.

A series of lessons learned from equipment failures in uranium processing facilities are summarized in Appendix III. Analysis of these equipment failures demonstrated a range of factors in the failures, including design, maintenance planning and execution, quality control, MOC, parts management, amongst others. Accordingly, these factors are also reviewed as part of the defect elimination programme. Investigating equipment failures is part of the continuous improvement process. However, there are many sources of continuous improvement, not only in the case of equipment failure, but including efficiency across all work management processes, shutdown management and production, as well as other internal processes which support maintenance and operations.

3.1.6. Allocation of resources

Any organization has a limited number of resources available to them, both financial and human, therefore it is critical to utilize the available resources in the most efficient manner to achieve the necessary deliverables of the organization. This is most often set and directed by senior management. Asset management needs to be considered as part of this resource allocation at the initiation of a mine or milling operation and cannot be left as an afterthought.

3.1.7. Deliverables (value)

An effective asset management framework delivers to the organization reliable equipment, planned availability of the equipment, a method to track and monitor asset related regulatory compliance, and minimum life cycle costs to the owner/operator.

This publication incorporates the principals for strategic asset management from Refs [13, 14, 26, 27].

3.2. LIFE CYCLE ANALYSIS

Asset life cycle analysis is a technical engineering review of the mechanical components, that as a group, make up a mine or processing facility. The purpose of this analysis is to align the economic business case of the facility with the equipment that is proposed to be used, in order to optimize profitability.

In the case of a uranium mine and processing facility, the owner/operator would consider the size of the ore body, and the expected time frame that the mine would be supplying ore to the processing facility. This is referred to as the 'life of mine'. For example, if the expected time frame to extract ore from a mine is 30 years, then the asset life cycle analysis would take into consideration that the assets are required to be reliably operational for the full extent of the 30 year time frame, or a strategy to ensure that equipment is replaced or refurbished as needed during that 30 year span. The analysis will also take into consideration how the equipment will be decommissioned and disposed of at the end of its useful life, as well as during closure works.

An optimized life cycle analysis would holistically consider the initial capital investment of designing and installing assets, as well as the routine maintenance costs associated with that design decision. For example, consider a large holding tank. It could be designed out of a carbon steel material with rubber lining that has a lower initial capital cost of $1 million, but is susceptible to high wear rate, or corrosion rate, and the rubber lining requires refurbishment every 5 years at a cost of $500 000, giving a total cost over 30 years of $1 million + (6 × $500 000) = $4 million. Alternatively, the holding tank could be designed out of a robust stainless steel material that is resistant to corrosion and wear, at an initial higher capital cost of $2 million, but only needs refurbishment after 15 years of service at a cost of $1 million, giving a total cost over 30 years of $2 million + $1 million = $3 million[3]. In this example, the higher initial cost comes with less maintenance cost, ending with a total life cycle cost that is lower, and therefore the preferred long term decision. However, in this scenario, the higher initial capital outlay may be prohibitive. Where capital outlay is available, the lower total life cycle cost option is ideally selected. This same life cycle analysis may be extended to all types of assets within a uranium mine and processing facility, to determine the preferred economics for a project.

During the design phase of a uranium mine and processing facility, maintenance engineers, design engineers, metallurgical process engineers, amongst others, are to be consulted when determining appropriate construction materials that meet the needs of all aspects of the process, including the type of fluids and environmental conditions that the asset is to be exposed to. This is critical for uranium processing operations, as these typically utilize chemicals such as sulphuric acid, ammonia, chlorides, organic kerosene and other reagents that may be significantly acidic or caustic, and/or abrasive materials which may wear or damage assets in unintended or unexpected ways. Uranium mines and processing facilities often differ to other mineral production facilities and other heavy industries, where the same equipment can be otherwise used without negative consequence.

Throughout the life cycle of the processing facility, newly discovered ore bodies in proximity may become feasible for extraction, thus extending the life of mine, or increasing production volumes. The owner/operator may regard this as an opportunity to extend the economic gain from investment in the original infrastructure. However, this production scenario may not have been considered in the original design or in the existing asset management strategy of the processing facility. In this case, further life cycle analysis would be required, beginning with an assessment of the current condition of the processing equipment. Capital expenditure may be required to replace or refurbish the assets so that they would be capable of processing the additional ore. For example, this may include replacing thinning pipelines, renovating building structures, relining process vessels, replacing holding tanks, and the many other considerations of the mechanical, electrical and structural components.

---

[3] Monetary values presented are provided as arbitrary figures to illustrate a quantitative assessment process.

3.3. ORGANIZATION STRUCTURE

Establishing and maintaining an organization structure for human resources is essential to have an accountable and effective asset management programme. A typical organization structure required to support the asset management for a uranium mine or uranium processing facility is provided in Fig. 5. Individuals performing in the roles require the specific knowledge, leadership abilities and technical skills for each position.

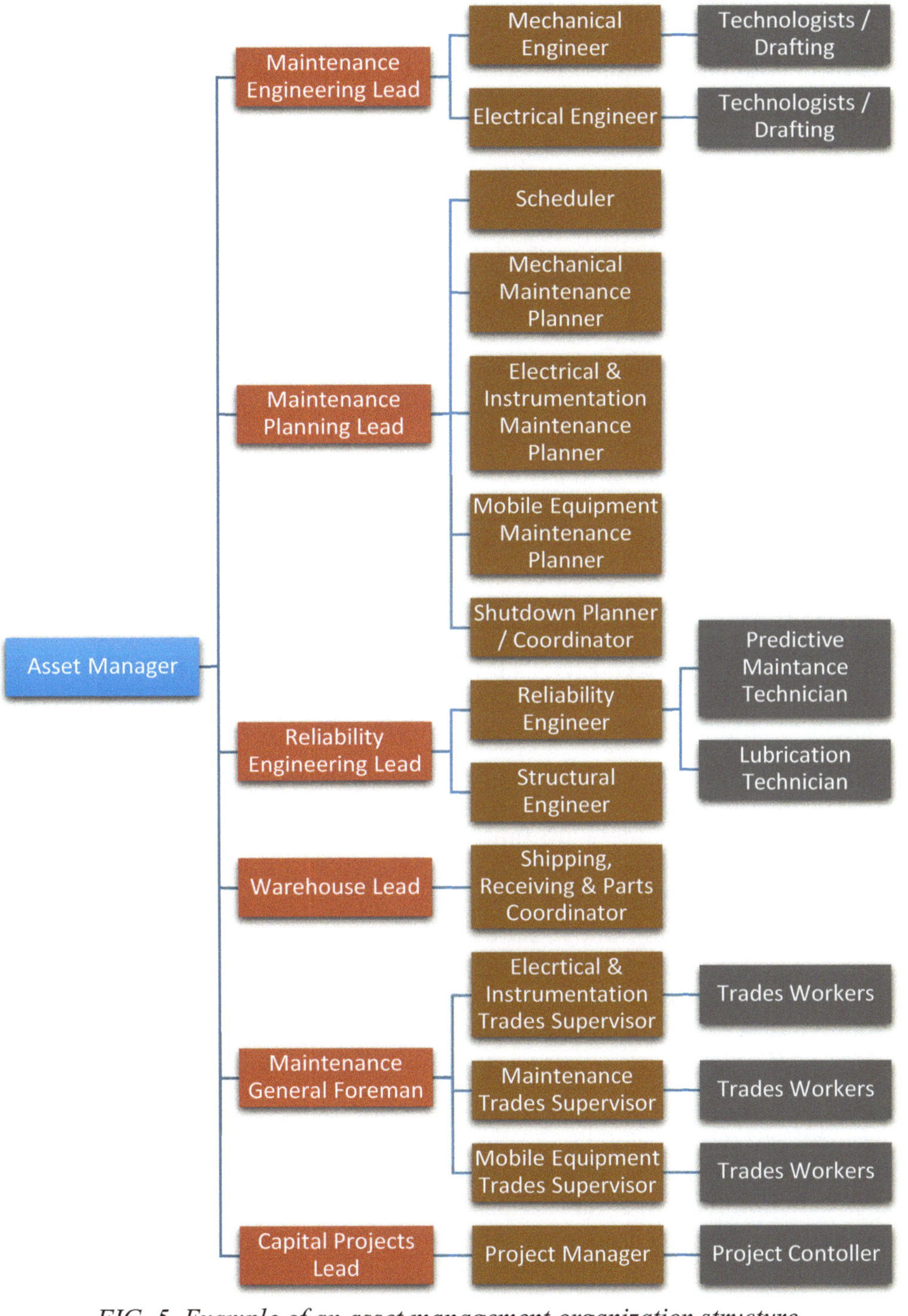

FIG. 5. Example of an asset management organization structure.

In addition to the positions shown in the organization structure in Fig. 5, there are typically other departments at the mine or processing facility providing support and services to the asset management group, such as production operations, occupational health and safety (including radiation safety), environment, accounting, human resources and training. If there are a large number of third party maintenance contractors utilized, then a separate role of 'contractor coordinator' is often employed to manage the contractor groups providing support to maintenance tradespersons' supervisors.

Generally, a maintenance tradespersons' supervisor would be tasked with overseeing a team (also called a crew) of 8 to 16 tradespersons. The supervisor will ideally be trained and qualified in the same area of maintenance work that their assigned tradespersons are tasked with.

The quantity of maintenance planner positions required for asset management is based on two interconnecting factors:

(a) The maturity of the 'plans' available on record for maintenance works;
(b) The number of maintenance trades workers that are on the maintenance team.

For example, if detailed plans are not available, then each maintenance task will require new planning. In this case, typically one planner can prepare plans for 8–10 maintenance tradespersons. Over time however, as an operation's asset management programme becomes more mature, a repository of maintenance plans will become available, especially in the case of repetitive maintenance works. This results in less planning effort to prepare for maintenance execution, and as such, a single planner can effectively complete planning for more tradespersons. When the plan repository is more mature, then one planner can typically prepare plans for 15–20 maintenance tradespersons. Supporting maintenance planners, are maintenance schedulers, who have responsibility for scheduling the works generated by planners. One maintenance scheduler supporting three to four maintenance planners, is a typical ratio.

4. ASSET MANAGEMENT PROGRAMME DEVELOPMENT

An asset is any equipment or infrastructure acquired for the mine or processing facility for the purpose of extraction or processing of ore (e.g. crushers, mills, tanks, infrastructure, haul trucks, etc.). An effective asset management programme needs to balance the cost, risk and performance of each asset. As there are finite resources available to perform maintenance, the most valuable use of these resources needs to be found.

Each piece of equipment therefore requires a unique asset management programme to be developed for it, which achieves a balance of cost, risk and performance. Failure of an asset is associated with its level of risk to occupational health, safety, the environment or production. Understanding the risk of failure associated with each piece of equipment is therefore a crucial first step to designing an effective asset management programme. By understanding the risk associated with the failure of each piece of equipment, a systematic, effective and cost effective asset management programme can be developed including the replacement and refurbishment strategy, type and frequency of monitoring, maintenance and spare parts to be kept in inventory. This strategy will minimize the risk to production, personnel and the environment, while optimizing maintenance and costs. This section focuses on the process to develop an asset management programme for a uranium processing facility, including developing a functional location structure for equipment, assessing the asset criticality, developing maintenance tactics and assessing the spare parts that are kept for assets.

4.1. ASSET REGISTER (EQUIPMENT HIERARCHY AND FUNCTIONAL LOCATION DEVELOPMENT)

All assets within a processing facility need to be recorded in an asset register and arranged in a hierarchy. This needs to organize assets in a way that allows them to be easily searched and shows the relationship between the levels of the hierarchy. This register can also be used to record other details, for example, the maintenance history or costs associated with the asset.

An example hierarchy provided in this publication has been adopted from Ref. [29], shown in Table 5. It includes an example of what may be included at each level of an equipment hierarchy which is the functional location of the asset (this is often referred to as the 'Floc').

Organizations with less complex operating sites, may simply structure the hierarchy using only levels 4–6, with levels 7–9 only being used as required.

Levels 1–5 provide information regarding the use and location within a mine or processing facility. Levels 6–9 are used to breakdown the equipment further.

The subunit (level 7 of the hierarchy) and maintainable item (level 8 of the hierarchy) are the individual components that make up the equipment unit. Typically, this includes components that can be changed out or replaced if they fail.

TABLE 5. EQUIPMENT HIERARCHY (ADAPTED FROM REF. [29])

Hierarchy			Examples
Use / Location	1	Industry	
	2	Organization or business category	
	3	Installation	Uranium production facility, water processing facility, mining facility
	4	Plant/unit	Mining equipment, primary crushing circuit, secondary crushing circuit, milling circuit, scrubbers, leaching, counter current decantation (CCD), flocculent plant, tailings and neutralization, clarification, solvent extraction, precipitation, calcination, product packing
	5	Section system	Individual trucks, individual leach tanks or CCD, individual mills, cyclones, conveyors, individual scrubber system in each area
Equipment subdivision	6	Equipment unit	Electric motor, engine, gearbox, compressor, generator, motor, heat exchange, tank, pump, piping
	7	Subunit	For a pump: power transmission, pump unit, control and monitoring, lubrication system
	8	Component/ maintainable item	For pump unit: casing, impellor, shaft, bearings, seals, valves, piping,
	9	Part	Individual parts of the maintainable item

It is worth noting the commonly used term 'productive unit', which is typically a 'section/system' asset (level 5 of the hierarchy). However, the productive unit is any asset which achieves something for the material (e.g. transports materials, transforms materials or provides storage for materials). Examples are individual haul trucks or bulldozers, ship unloaders, major pipelines, mills, silos and tanks.

A defined equipment hierarchy is important as it is required to complete ACAs. Asset criticality assessments are always performed on the 'equipment unit' (level 6 of the hierarchy). However, the following exceptions can apply when completing an ACA at the section/system level (level 5):

— Asset criticality assessments may be performed at the section/system if it comprises a fleet or class of assets with similar inherent risk profiles that an operation intends to manage using a common asset management strategy. An example of this may be a haul truck fleet or multiple similar mills, crushers or tanks performing the same service.
— Continuous productive units, such as a water pipeline or railway which may be segregated into separate higher criticality segments, for example, at bridges or high traffic areas.
— Asset criticality assessments may be performed at the section/system level if the section/system has a higher criticality as a whole when compared with an individual productive unit. For example, two pumps in a duty/standby arrangement may individually have lower criticalities than the combined pumping system.

4.2. ASSET CRITICALITY ASSESSMENT

Understanding asset criticality helps to prioritize allocation of an organization's resources and budgets for assets, for the purpose of maximizing equipment availability and reducing the risk of unplanned downtime.

An ACA utilizes the risk assessment process outlined in Section 2.5 to categorize the risk level and criticality of the asset within the context of the organization's objectives. Critical assets are identified through a process of assessing equipment failure modes, the consequence of a failure and the likelihood of a failure, in order to establish the risk level. These concepts are described in more details in the following subsections.

Once an operation's overall asset criticality is understood, each piece of equipment can be maintained in accordance with an asset management strategy that corresponds with the risk category (e.g. from low risk to critical risk assets). The process of categorizing assets based on risk is the basis for developing maintenance tactics for equipment. Further information on developing maintenance tactics can be found in Section 4.3.

4.2.1. Assess productive unit functions and failure modes

Once the equipment hierarchy is established, the main function of each equipment unit needs to be identified along with the potential failure modes associated with the equipment unit. At this point, the MRO is used for each failure mode, which is the maximum consequence which could reasonably be expected to occur. For example, the primary function of a tank is to contain a liquid and therefore the failure mode with the highest MRO for a large tank would be a rupture of the tank. For a crusher, its primary purpose is to crush rock and the highest MRO for this would be if it could not operate due to failure of a major component or worn liners. One thing to note is that the failure modes are not a reflection of the current condition of the equipment.

The main function and failure modes need to be determined by an assessment team with knowledge of equipment failure scenarios and understanding of the risks. An asset may have multiple failure modes that can prevent it from performing its primary function. If required, multiple failure scenarios can be assessed to identify the failure scenario with the highest MRO.

4.2.2. Assess consequence of failure scenario

To assess the consequence of any equipment failure scenario, it is necessary to first understand how the equipment fails. Equipment failures can be categorized into the following:

— Operational based failure;
— Induced failures;
— Intermittent failures;
— Wear out failures.

Operational based failures occur by simply operating an asset outside of its original design intent.

Induced failures may be described as a failure mode external to the asset that causes the failure. An example of this is a misaligned coupling that results in a bearing failure within the asset. This type of failure is often identified by processing monitoring and predictive monitoring strategies, in combination with further analysis to determine the root cause of the failure and eliminate induced failures.

Intermittent failures are those types of failures that may happen at any time and can also be referred to as random failures. These types of failures make it difficult or impossible to determine the 'mean time between failure'. The best strategy to detect these types of failures is through process and predictive monitoring. Past practice has shown that preventive maintenance strategies are ineffective at detecting intermittent failures and may in fact be a root cause of such failures.

Wear out failures have a known mean time between failure, based on past operational practice or original equipment manufacturer (OEM) specifications. Time based refurbishment has been shown to be the most effective maintenance strategy for this type of failure. Predictive maintenance monitoring may also be used as a strategy to identify wear out failures.

For any given equipment failure scenario, the maximum reasonable consequence needs to be assessed, using the organization's risk assessment consequence descriptors. A description of the assessment of consequence, descriptors, and examples for uranium mining and processing facilities, are provided in Section 2.5.1 and Table 2.

One method to improve the accuracy of an ACA is to perform the equivalent of a quantitative assessment of the consequence to determine the financial loss to the organization as a result of the equipment failure. This financial loss does not take into consideration the consequences of occupational health, safety, environment, community, reputation, etc., to the organization. The basis for quantitative assessment is cost, which is informed by a description of the event and the damage or duration of outage. This cost can then be checked against the organization's consequence descriptors for financial loss. An example of the consequence of a pump failure is outlined in Fig. 6, assuming there are no control measures in place.

Event: Pump unable to achieve required flow due to worn impeller.

Damage: 3 days downtime

Assume 1 hour down time = \$65 000 lost revenue

Total cost = Lost Revenue + Labour Cost + Parts Cost

= 3 days × 24 hrs × (\$65 000/hour revenue) + (3 × 8 × \$80/hour labour cost) + \$6400 parts cost
= \$4 688 320 [(4)]

FIG. 6. Example of the financial loss as a consequence of a pump failure.

4.2.3. Assess likelihood of failure scenario

For any given equipment failure scenario, the maximum reasonable likelihood of that scenario occurring is assessed using the organization's risk assessment likelihood descriptors. A description of the assessment of likelihood, descriptors, and examples for uranium mining and processing facilities, are provided in Section 2.5.2 and Table 3. When assessing failures of equipment, the assessment for the likelihood of failure needs to be made by technical experts who have a thorough understanding of damage mechanisms of the equipment degradation including when and how failure occurs. It is important for the scenario being assessed to be credible.

4.2.4. Assess risk category

For any given equipment failure scenario, the final step in assessing asset criticality is to categorize the level of risk, using the organization's risk assessment descriptors. A description of the assessment of risk and risk descriptors are provided in Section 2.5.3 and Table 4. Typically, critical assets will be those categorized with a risk level of 'critical' or 'high', where failure will result in a significant occupational health, safety or environmental consequences, or an interruption of production, or an underperformance of the asset.

Once asset criticality is established, each piece of equipment can be maintained in accordance with a corresponding asset management strategy. For higher risk assets, the risk may be mitigated through a variety of maintenance strategies.

By establishing the maintenance strategy and developing the maintenance tactics, the likelihood and consequence of the failure scenario will generally be reduced. This revised risk is known as the residual risk. It is important to note that the maintenance strategies will reduce the risk associated with the equipment; however, the original criticality rating for each piece of equipment prior to maintenance strategies being implemented remains unchanged.

4.3. MAINTENANCE TACTICS DEVELOPMENT

Maintenance tactics refers to the strategy that governs how equipment will be maintained. The tactics for maintenance are best identified and considered at the design stage of a uranium mine

[4] Monetary values presented are provided as arbitrary figures to illustrate a quantitative assessment process.

and processing facility. Equipment is expected to be designed so that the specified maintenance tactics can be executed in the most efficient manner. For instance, if a pump is to be installed as part of the facility design, consideration needs to be given to how to safely remove the pump so it can be replaced or maintained. This may include strategically placing valves in the pipe works, installing a lifting device above the pump, as well as a safe method to move the pump out of the installation and to a location where it can be refurbished.

There are four primary strategic maintenance approaches, starting from simplest to advanced:

(1) Run to failure: Simply continue to operate the equipment until it fails. This is a reasonable strategy if the failure has low consequences or no impact, such as choosing to replace light bulbs when they burn out. For run to fail strategies within the processing plant, there are usually duty/standby arrangements, spares in stock or replacement equipment is able to be procured with short lead times and has simple replacement processes.
(2) Preventive maintenance: Replace or rebuild the equipment on a planned calendar based frequency before it fails. This strategy typically means that the useful life of the equipment is not fully realized, and maintenance costs may be higher overall than compared to more advanced strategies, such as replacing equipment based on run time hours or a condition based assessment.
(3) Predictive maintenance: Uses technology to analyse current condition of equipment during an inspection in order to identify early stages of failures, and then trigger a replacement or refurbishment before complete failure occurs. This strategy would employ a specialist to perform vibration monitoring, lubrication analysis, infrared inspection, and/or sonic inspection on a routine basis in order to trend indicators of equipment early stage failure, such as on a mill pump, gearbox or electric motor. This strategy is typically utilized for equipment which has been assessed as critical to the process.
(4) Proactive/strategic maintenance: Uses computer systems to track, trend and analyse a large quantity of data inputs to identify early signs of equipment malperformance, and then triggers a notice to the maintenance team to perform mitigating and corrective action. This is an adaptive artificial intelligence learning system that continually improves itself and is capable of making rational decisions without human input.

The senior management of an organization will develop an economic strategy to determine planned downtime of the facility in order to execute maintenance tasks that require plant or equipment outages. This strategy will take into consideration many factors and may include budgets available for maintenance, the production requirements, market forecasts, contract requirements, plant age, social requirements, capital requirements, amongst many other factors. Once the production plan has been determined for the year and the target for process availability had been set, such as 85–90% operating time, the remaining time (i.e. 15% in this case) would be allocated for planned major maintenance shutdowns, smaller outages and routine maintenance. For example, this may result in an annual production plan allowing for 310 operating days per year (i.e. 85% operating time) and the balance being allocated for planned and scheduled shutdown for maintenance tasks. During these planned outages, the maintenance team would execute major maintenance activities and inspections that ensure the equipment can operate safely and efficiently for the remaining 310 days of planned processing operations.

To effectively execute a planned maintenance outage, the operations and maintenance activities are required to be planned, scheduled, and integrated. Good practice is for a processing operations specialist to develop shutdown and startup strategies and procedures, which would include cleaning procedures, lock out, and isolations so that the maintenance workers can best utilize the time assigned to them to execute the tasks that can only be accomplished in the short shutdown time allotted.

Annual budgeting for the facility needs align with the maintenance tactics put in place. This means that human resource planning, vacation scheduling, long lead time purchased items, engineering design work, etc., will be considered and integrated into operating plans. If contractors are used to perform specialty maintenance activities, they need to be identified, and negotiated contracts in place well ahead of when the work is expected to be needed. During planned maintenance shutdowns, the number of workers on site may be much larger than during normal operations, so additional logistical considerations include the provision of specific safety equipment and training required for the tasks. If the site is in a remote location, further considerations may include logistics considerations (e.g. transport, accommodation, meals, and office space).

4.4. FAILURE MODES AND EFFECTS ANALYSIS (FMEA)

Failure modes and effects analysis (FMEA) is a process to develop maintenance tactics for individual assets from first principles. FMEA is a method designed to:

— Identify and fully understand potential failure modes and their causes, and the effects of failure on the system or end users, for a given product or process;
— Assess the risk associated with the identified failure modes, effects and causes, and prioritize issues for corrective action;
— Identify and carry out corrective actions to address the most serious concerns.

To conduct an FMEA, the equipment is subdivided into its various components. The failure modes and effects of each 'component' (level 8 of the equipment hierarchy) are analysed. An example of the functional location for a mill pump, using the equipment hierarchy, is shown in Table 6. In this example, the analysis would be performed for the impellor component.

TABLE 6. EXAMPLE OF AN EQUIPMENT HIERARCHY TO DETERMINE FAILURE MODE (ADAPTED FROM REF. [29])

Hierarchy			Examples
Use / Location	1	Industry	
	2	Organization or business category	
	3	Installation	Uranium production facility, water processing facility, mining facility
	4	Plant/unit	Milling circuit
	5	Section/system	Mill pumping system
Equipment subdivision	6	Equipment unit	Mill Pump #1
	7	Sub-unit	Pump unit
	8	Component/maintainable item	Impellor
	9	Functional failure	Flowrate below required level
	10	Failure mode	Impellor worn

The following section outlines the steps to conduct an FMEA for the purpose of developing asset management activities.

1. Function;
2. Functional failure;
3. Failure mode;
4. Failure pattern;
5. Failure effect;
6. Failure priority;
7. Proactive tasks.

Descriptors of commonly used terminology within the FMEA are provided below, using the mill pumping system referred to in Table 6, as an example for performing the FMEA.

4.4.1. Function

The first step of an FMEA is to understand and identify the functions of the assets. Assets are acquired by every organization to perform certain functions. The value delivered by the functions and associated desired standards of performance of assets are based on the operational requirements of the organization. These functions are divided into two main categories, 'primary functions' and 'secondary functions, described in more detail below.

(1) Primary functions: The primary function is what the owner/operator of a physical asset or system wants it to do. In the case of the mill pump example, the primary function of the pump is to pump slurry from point A to point B at a certain rate (e.g. 300 m^3/hour) and at a certain pressure (e.g. 400 kPa).

(2) Secondary Functions (E,S,C,A,P,E,S acronym): Secondary functions are auxiliary functions in addition to the primary function. Secondary functions include:
- E: Environmental;
- S: Safety;
- C: Containment/contamination;
- A: Appearance;
- P: Protection — systems in place to prevent and protect;
- E: Economy/efficiency;
- S: Superfluous.

Whilst the Secondary functions are important, these will not be addressed specifically in this example of the mill pump.

4.4.2. Functional failure

The second step in the FMEA process is to define the functional failures of the asset. A functional failure or fault according to Ref. [29] is "a state of an item characterized by inability to perform a required function, excluding such inability during preventive maintenance or other planned actions, or due to lack of external resources".

An understanding of the various ways in which an equipment fault or failure may prevent it from fulfilling its function will inform the analysis. In the mill pump example, as long as the pump is delivering the designed flow and at the prescribed operating pressure then it is meeting its defined performance criteria. However, as soon as the performance (i.e. flow) drops below a predetermined rate then a functional failure has occurred even though the pump is still operating.

4.4.3. Failure mode

Once the functional failure condition is identified, the next step in the FMEA process is to specify the failure mode, which is defined as the observed manner of failure. Failure modes are normally related to the 'equipment unit' (Level 6 in Table 6) in the hierarchy. The failure modes can be categorized in three types:

- The desired function is not obtained (e.g. fail to start);
- There is a deviation in a specified function outside accepted limits (e.g. reduced output);
- There is a failure indication observed, but there is no immediate and critical impact on equipment unit function (e.g. leakage).

The failure mode describes the failure indication at the equipment unit level, while the failure descriptor describes the cause of failure on the lowest level within the equipment hierarchy for which this information is known. It is important within the FMEA to understand how the equipment fails and what causes each functional failure. To illustrate this using the mill pump example, the failure mode may be a worn impeller due to medium abrasion from the material being pumped.

4.4.4. Failure pattern

The failure pattern of the asset is used to specify details on underlying and direct causes of failure and assist in determining the most appropriate maintenance tactic to apply. Common failure patterns include the following:

- Infant mortality: This failure is early in the life of the asset typically due to manufacturing defects, design issues, installation or commissioning defects.
- Random: This failure type is low and constant over the life of the asset (i.e. weather or operational misuse).
- Wear out: The failure rate increases significantly at the end of life due to wear and age of the equipment.
- Bathtub: This is a common combination of infant mortality, random and wear out, where there is a higher likelihood of failure at startup and at the end of life.
- Fatigue: Where the possibility of failure consistently increases with the age of the equipment.
- Initial break in period: The likelihood of failure starts low and then rises to a constant level.

In the case of the mill pump example, a failure pattern of wear out resulting in a failure effect of an overflowing mill discharge sump, will be utilized to illustrate the FMEA process.

4.4.5. Failure effect

The failure effect step involves detailing the consequences of each potential failure mode on the system, component, or process being analysed. The failure effect description articulates how the failure impacts the operation, functionality, performance, safety, and/or regulatory compliance of the product or process. Failure effects can also include the downtime required to reinstate the equipment to the required standard or cost associated with the refurbishment or replacement of the equipment.

Generally, there are four main categories of consequences of failure:

— Hidden: Where the failure effects are not evident to the operating team under normal circumstances;
— Safety and environmental: Where the failure effects constitute a risk to safety or the environment;
— Operational: Where the failure leads to a loss of operational capability;
— Non-operational: Where the only consequences of the failure are the direct cost of the repair.

4.4.6. Priority

To prioritize resources for maintenance tactics, it is helpful to use a quantitative approach to assess the severity, occurrence and detection ability of the failure. The current controls are outlined at this point (if completing a review of tactics).

The organization will need to establish an assessment tool, which described categories and ranking methods to assess severity, occurrence and detection. The following provides an example of category and ranking descriptors:

— Severity (S): Impact of the failure on safety, environment and operational performance, rated from 1 (least severe) to 10 (most severe);
— Occurrence (O): Likelihood of the failure occurring, rated from 1 (rare) to 10 (frequent);
— Detection (D): Ability of current controls to detect the failure before it has a significant impact, rated from 1 (easy to detect) to 10 (hard to detect).

Furthermore, the organization will need to establish a method of quantitative assessment to establish the risk priority. A commonly used measure of risk priority can be obtained from the product of severity, occurrence and detection to establish a risk priority number (RPN), as shown in the equation below:

$$RPN = S \times O \times D \quad (1)$$

The risk priority number scores can be used to prioritize which failure modes need the most immediate attention. Where a higher RPN indicates a higher priority for action.

4.4.7. Proactive tasks

The final step in the FMEA process is to develop tactics, or maintenance tasks, to predict or prevent failures identified through the analysis outlined in the steps above.

An example of a completed FMEA for a pump is shown in Table 7 below. The analysis provides direction on the failure mode and interval of failure such that maintenance tactics can be developed to prevent these failures from occurring, in line with the strategy of the organization (i.e. what component to inspect, change out, refurbish, etc.).

The FMEA process lends itself well to the development of maintenance tactics required for rotating assets. A series of examples of FMEA assessments for rotating equipment is also included in Appendix IV for conveyors, gearboxes and couplings. The tables are provided as industry examples of those used by reliability and maintenance engineers to assist with developing maintenance tactics and ensure optimal equipment performance.

TABLE 7. PUMP FMEA

Function	Functional Failure	Failure Mode	Failure Cause	Effects of Failure	Failure Pattern	RPN Number (S×O×D)	Maintenance Actions	Frequency
To move fluid	Inability to pump fluid	Bearing failure	Wear due to overuse	Overheating, increased vibration, pump stops	Early	504 (8×9×7)	Inspect and replace bearings	Every 6 months
			Lack of lubrication	Increased friction, overheating	Random	280 (7×8×5)	Lubricate bearings	Every 3 months
		Impeller wear	Abrasive particles	Reduced efficiency, increased vibration	Wear-out	360 (6×6×10)	Inspect and replace impeller	Annually
			Corrosion	Impeller degradation, reduced flow rate	Wear-out	270 (9×6×5)	Regular cleaning, anti-corrosion coating	Every 2 years
		Seal leakage	Improper installation	Fluid leakage, environmental hazard	Random	630 (9×10×7)	Inspect and replace seals	Every 6 months
			Wear and tear	Fluid leakage, decreased pressure	Wear-out	560 (8×10×7)	Regular inspection, replace as needed	Annually
		Motor failure	Electrical fault	Pump stops, possible electrical hazard	Random	450 (9×5×10)	Electrical inspection, replace motor	As needed
			Overloading	Overheating, reduced lifespan of pump	Random	320 (8×8×5)	Monitor load, install protection devices	Every 3 months
		Electrical failure	Power surge	Immediate stop, potential damage to components	Random	200 (5×8×5)	Install surge protectors	After every surge event
			Corrosion of wiring	Intermittent or total power failure	Wear-out	225 (5×9×5)	Regular inspection, replace wiring	Every 2 years
To maintain pressure	Loss of pressure	Clogged filter	Contaminated fluid	Reduced flow rate, pump overload	Increasing	720 (8×9×10)	Clean or replace filters	Monthly
		Leaking discharge valve	Wear or damage	Reduced system pressure, energy wastage	Wear-out	400 (8×5×10)	Inspect and replace valve	Annually

4.5. FIXED CRITICAL ASSETS

Fixed critical assets are static, non-rotating assets that require maintenance and inspections regimes to be in place to ensure availability, safety of personnel and protection of the environment. Examples of fixed critical assets include the following:

- — Pressure vessels;
- — Storage tanks;
- — Piping systems;
- — Pressure relief valves (or pressure safety valves/burst disks);
- — Cranes;
- — Structures;
- — Ducting;
- — Electrical equipment (including variable frequency drives and motor control centres).

Due to the static nature of these assets and less frequent ongoing maintenance compared to rotating equipment, they may, incorrectly, be overlooked when developing asset management programmes. Large pressure vessels, tanks and piping systems can have significant safety risks associated with them. Catastrophic failure of these assets is often attributed to major process safety incidents and multiple fatality incidents. There may also be significant production consequences if failure of these assets does occur.

Fixed assets are to be included in ACAs (following the same process outlined in Section 4.2) to prioritize and develop maintenance tactics. These assets may also have statutory or regulatory requirements associated with them which need to be fulfilled.

The maintenance tactics required to maintain fixed assets differs from rotating assets and often requires asset integrity engineers to oversee their management. The maintenance tactics are heavily reliant on visual inspections and non-intrusive inspection methods such as ultrasonic thickness (UT) testing. More detail on maintaining these assets is provided in Section 6.

A tool that can be used to better understand the risks and controls associated with fixed critical assets, primarily piping, pressure vessels and tanks, is through a risk based inspection (RBI) process. The RBI process is similar to the FMEA process described in Section 4.4, however it was developed by the oil and gas industry with the aim to more closely understand the damage mechanisms and failure modes. This process is outlined in Appendix V.

An RBI programme is time consuming and costly to produce and needs to be used selectively for the highest risk or high maintenance cost assets only. For all other fixed assets, the ACA process and FMEA process is appropriate.

4.6. SPARE PARTS MANAGEMENT

Maintenance programmes are heavily reliant on having spare parts available or being able to procure the spares parts in an appropriate time frame. A spare parts strategy will need to be developed that is aligned with the criticality for the equipment and its maintenance strategy.

An increased quantity of spares may assist to increase plant availability through fast change out of failed equipment. However, the cost of spares (as inventory carrying costs) and space required to properly store parts, needs to be balanced against the organization’s operating strategy, including planned equipment availability and cash flow requirements. Good practice in life cycle asset management is to have spares being used at least once per year for consumables. Other critical spares that are rarely changed out but have long delivery times ought to be kept on hand (e.g. ring gear for a grinding mill, due to long lead manufacturing times).

The subsection below provides a high level overview of the reasons that spares are kept at the location of operations. This section provides guidance on spares criticality assessments and also includes warehousing considerations for spares once they have been determined as being required. Spares are often poorly managed in a warehouse including being left outside or exposed to the elements, poor packaging, poor racking and storage, and also poorly maintained. These scenarios can lead to spare parts being damaged or in poor condition when installed, often with additional avoidable cost incurred by the operation.

4.6.1. Spares classifications

There are multiple classifications for spares depending on how they are procured, managed and accounted for during their life cycle. While not exhaustive, the following section describes the classification of, and management, of key spares.

4.6.1.1. Rotable and repairable spares

Some spare parts are classified as rotable (also known as serviceable or replacement units). These are parts that wear down during their service life but can be restored to a functional condition, sometime repeatably. The worn part is serviced or repaired after being removed from service and stored in the warehouse until required. These spares need to be tracked throughout their life cycle as individual equipment items to ensure their maintenance history is understood. This is particularly important where refurbishment is not completed each time the component is rotated. Rotable items are usually high cost components that can be repaired or refurbished at a lower cost relative to replacement. Typical examples are pumps, gearboxes and large electric motors. Good practice is to track the repair history of all rotable spares, with the intent to determine the costs associated with each spare. If the repair costs exceed the repurchase costs the rotable asset is expected to be replaced.

4.6.1.2. Registered equipment spares

Spares of registered equipment need to be uniquely identified to ensure they comply with national and/or local statutory requirements, usually regarding inspection or testing intervals and records tracking. Typical examples include PRVs and PSVs.

4.6.1.3. Insurance spares

Insurance spares are those selected for production critical equipment, where a spare is required to minimize downtime. Insurance spares can also be selected to mitigate the potential impacts on the environment, and on the safety of workers and the public, in the event of a failure. The spares assessment for these assets is primarily dependent on the lead time from suppliers and less dependent on factors like cost and warehousing. Insurance spares can vary in size and function from large pieces of mechanical equipment, like mill bearings and gearboxes, to small electronic components like control system cards. Insurance spares may never be used during the life cycle on an asset, thus the prior history of use of spare parts is not taken into consideration to determine a need to keep these parts in stock.

4.6.2. Spares criticality assessment

The decision making process to determine if a spare part (or quantity of spare parts) is required to be kept takes into consideration the equipment criticality assessment (see Section 4.2), failure detection monitoring (e.g. predictive maintenance and condition monitoring), physical size of the spare, warehousing costs, delivery time and cost of the spare.

In general, if the time between detection of failure and catastrophic component failure is less than the lead time for delivery and there is no redundancy in the design, then the component will need to be held in stock as production, occupational health, safety, or the environment may be impacted if failures occur. Other considerations include whether the equipment can be repaired or bypassed until the new equipment is sourced or whether additional monitoring is needed (if reduced output has occurred). Additional factors to be considered in the spares criticality assessment process include:

— Production impact;
— Mean time before failure;
— Mean time to repair;
— Environmental impact;
— Occupational health and safety impact;
— Utilization and availability;
— Lead time to supply;
— Warehouse preservation effort;
— Capital expenditure.

4.6.3. General warehousing considerations

Equipment manufacturer specifications need to be reviewed for information on storage and maintenance requirement specific to that equipment. The following subsections provide guidance on storage and maintenance practices for spare parts. The shelf life of spare parts will vary depending on the type of packaging, preservative material, storage environment and maintenance during storage.

4.6.3.1. General

Examples of general good practices for warehousing are outlined below:

— Stored on bunded (or 'bermed') pallets, under cover, in a dry and cool location and in ventilated areas.
— Smaller equipment (e.g. up to 80 kg) strapped to pallets.
— Larger equipment stored on engineered and safety rated storage frames.
— Conveyor belts stood upright (i.e. not on their side) on a spindle frame and safely blocked to prevent movement.
— Electrical components stored on shelves or in cabinets that are electrically grounded.
— Silica gel adsorbent used to remove traces of moisture.
— Breather ports plugged with steel plugs, and the breathers attached to the gearbox in a clean plastic bag with a label indicating 'Attach to gearbox after installation'.
— Where specified, equipment to be empty of oil and tagged with a label indicating 'No oil during storage'.
— 'Repairable spares' label attached to the relevant equipment.
— Internal contents sealed and protected from the environment.

- Condition of the packaging checked every six months.
- Electrical components packaged using a vacuum seal process.

4.6.3.2. Batteries

Batteries in storage can be hazardous as they can give electrical charge and can leak or spill hazardous substances. There are a range of batteries on the market and care is needed to store batteries correctly. Storage is to be in accordance with manufacturer equipment specifications; however, the following general considerations apply to the storage of batteries:

- Stored away from potential ignition sources (e.g. heat, sparks, open flames, welding areas).
- Stored under cover, in a dry and cool location that is well ventilated.
- Stored on bunded pallets in case of a chemical spill from the battery.
- Stored on the bottom shelf may be considered so as to not mix battery chemicals with other chemicals.
- Batteries that are not encased in plastic or hard rubber are stored off concrete floors.
- Manufacturer safety data sheets are provided in the vicinity of storage.
- Emergency response equipment (e.g. fire suppression, spill kits) is available in the vicinity.

4.6.3.3. Rotating equipment

Examples of rotating equipment include pumps, drives, variable speed drives, compressors, fans, blowers, etc. Storage and practices for rotating equipment spares include the following:

- Stored in a vibration free environment (to prevent brinelling of bearings).
- Machined surfaces covered and maintained with an anti-corrosive agent or material.
- Exposed metal external to the item is protected from the environment to prevent corrosion.
- Shafts for coupling connections are covered with grease or greased fabric tape.
- Unsealed bearings are pre-packed with grease, and where there is absence of oil all internals are sprayed with a rust inhibiting lubricant (i.e. oil soluble).
- Bearing housings that are oil lubricated (but not force-fed) are filled with a rust preventing concentrate (with the oil level up to the bottom of the shaft).
- Bearing houses that are force-fed have the upper bearing cap and bearings removed.
- Rotating components are rotated every three months based on OEM specifications. Rotation ensures free movement, changes the load on the bearings to prevent false brinelling and ensures there is a lubrication film on the bearing surfaces. Each rotation is documented and tracked in the CMMS.

4.6.3.4. *Bearings*

Bearings condition is a critical factor in the reliable operation of rotating equipment, therefore they are to be stored correctly and maintained appropriately. The following are general good practices for storage of bearings:

— Stored horizontally and wrapped in plastic to prevent dust and moisture ingress.
— Stored in a vibration free environment (to prevent brinelling of bearings).
— White metal bearing assemblies (i.e. friction/plane bearings) are filled with oil as per the manufacturer's specifications. Oil rings and inspection covers are installed and the unit is rotated every three months.
— The bearing inner rings are rotated every six months, based on manufacturer's specifications. Each rotation is to be documented and tracked in the CMMS.

Long term storage (e.g. over five years) of capped (i.e. sealed or shielded) bearings is generally not possibly due to potential failure modes including:

— Grease separation in pre-greased (double shielded or sealed bearings);
— Grease settling to one side of the bearing;
— False brinelling damage;
— Corrosion.

4.6.3.5. *Motors*

Due to the dependence on material and process fluid movement within a processing facility and the potential for failure of motors, it is critical that motors are stored correctly and maintained while in storage. Key points on the storage of motors are described below:

— Larger motors with installed heaters have their heaters connected and tested monthly. This assists to prevent condensation forming inside the motor.
— Drive shafts are rotated monthly with the keyway stopped in a different position each month. It is good practice to stop the keyway at 130 degree increments, (i.e. 12 o'clock, 5 o'clock, 10 o'clock, 4 o'clock positions) on a monthly frequency. This ensures bearings remain lubricated and rested in different positions. Large drives above 2000 hp/1500 kW may need more frequent rotation to prevent the shaft bowing under the weight of the rotor.
— A 'motor storage card' is attached to the motor to help track rotation and other storage information (see example in Fig. 7).

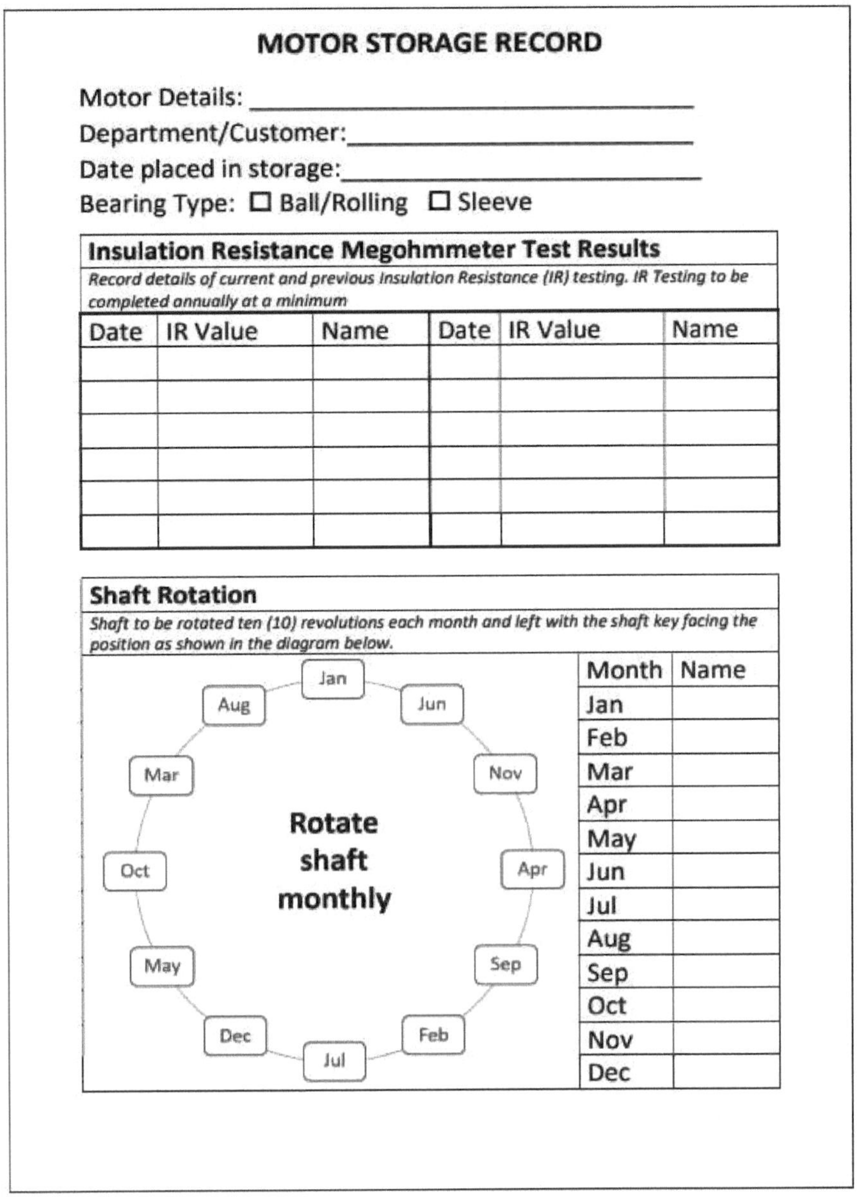

MOTOR STORAGE RECORD

Motor Details: ______________________________

Department/Customer: ______________________________

Date placed in storage: ______________________________

Bearing Type: ☐ Ball/Rolling ☐ Sleeve

Insulation Resistance Megohmmeter Test Results					
Record details of current and previous Insulation Resistance (IR) testing. IR Testing to be completed annually at a minimum					
Date	IR Value	Name	Date	IR Value	Name

Shaft Rotation

Shaft to be rotated ten (10) revolutions each month and left with the shaft key facing the position as shown in the diagram below.

Month	Name
Jan	
Feb	
Mar	
Apr	
May	
Jun	
Jul	
Aug	
Sep	
Oct	
Nov	
Dec	

FIG. 7. Example of card to track bearing rotation.

4.7. REGISTERED ASSETS

In many States, certain assets are required to be registered with a governing body. This is required to ensure the correct design, construction, inspection, maintenance and certification processes are in place to protect the safety of workers and the public. Registered assets are typically high risk equipment with potential for significant occupational health, safety, environmental and broader community consequences if not managed correctly. Requirements for registered assets will differ between Member States, however this will usually include equipment such as boilers, cranes, hoists, lifts or pressure equipment. Inspections are usually mandatory for registered equipment and in some cases authorized inspection authorities may be required by regulation.

The owner/operator of a uranium mine and processing facility is expected to maintain a registered asset database, ensuring that it captures key information relating to the equipment. This can be an electronic file or be within the CMMS. A registered equipment database contains at a minimum, the following:

— Functional location;
— Location description;
— Plant type and description;
— Manufacturer;
— Model number;
— Serial number;
— Equipment design registration number;
— Equipment design registration certification number or third party design verification number;
— Individual plant registration number (if applicable);
— Capacity/volume (for pressure vessels);
— Design pressure (for pressure vessels);
— Hazard level (for pressure vessels);
— Primary safe working load (for elevating work platforms, cranes and hoists);
— Secondary safe working load (for cranes and hoists);
— Commissioning date;
— Date of last statutory inspection;
— Statutory inspection period;
— Date of next statutory inspection.

5. MAINTENANCE PROGRAMME EXECUTION

Maintenance execution encompasses the process of completing work within the maintenance system, to ensure that maintenance is performed using the right process, resources and tools, so that it is completed efficiently and on time. This section details the processes and governance for asset reliability, the steps of a maintenance work management process and performance management as part of continuous improvement within uranium mines and processing facilities.

5.1. RELIABILITY

Reliability performance is defined according to Ref. [29] as the ability of an item to perform a required function under given conditions for a given time interval. This section will describe reliability considerations, commencing at the time of the design phase of a uranium mine and processing facility and throughout operations, and provides examples of reliability engineering processes.

Reliability, in an asset management context, enables measurement of an asset performing the required tasks successfully. The following metrics may be used for measuring equipment reliability performance:

— Availability;
— Mean time to repair;
— Mean time between failure;
— Mean time to failure after primary maintenance.

5.1.1. Design for reliability

When designing a uranium mine and processing facility for reliability, it is necessary to define the functional specifications for each asset within the processing facility (e.g. carry X tonnes for X distance at X speed, or pump slurry at X flow rate). In addition, the chemical nature of the materials that will be processed or stored by the assets also needs to be considered. These specifications form the basis from which the engineering and procurement teams design or select the asset.

The development of functional specifications is a formal, integrated design and procurement process which is based on the organization's requirements as specified in the operations' asset management strategy. The functional specifications ought to be detailed, clear, and appropriately cover performance, cost and risk requirements where relevant.

The organization needs to apply engineering principles (mechanical, chemical, electrical, structural, civil, process, mining, etc.), design standards and criteria to each asset. The purpose is to ensure the design, selection and construction of the asset meets its functional specifications. Examples would include bucket or tray sizes, materials selection to resist corrosion or design load cases. All relevant engineering expertise ought to be sought and incorporated to ensure a rigorous process. Relevant factors for consideration are:

- Battery/scope limits;
- Geography and climate;
- Regulatory conditions;
- Occupational health, safety, environment and community settings;
- Geological factors;
- Mining, processing and infrastructure considerations.

The safety in design principle supports inherently safer outcomes through the life cycle of an asset, by removing hazards in the early stages or through use of hierarchy of control to reduce risks, during the design process. This approach is industry good practice when compared to adding protective measures after asset construction, which often relies on administrative controls or PPE to reduce risk. For example, designing plant to include ventilation systems or scrubbers in uranium ore crushing areas is preferred, when compared to relying on workers to wear personal respiratory protective equipment.

Designs need to comply with the requirements of the organization's asset management strategy. The asset criticality level will determine how rigorously the strategy needs to be applied. For critical assets, operational procedures and maintenance tactics strongly inform the design process. Examples include consideration of the plant shutdown schedule, requirements to train production operators, required tooling, etc. The design process needs to also consider the compatibility between maintenance tactics and spares.

For critical assets, the operation needs to be able to demonstrate both reliability engineering and life cycle cost modelling of various designs or configurations.

The design process also needs to identify and comply to relevant internal and external standards (e.g. international standards, national guidelines, industry standards).

5.1.2. Operate for reliability

The operations personnel of a uranium mine and processing facility are key parties in the performance, reliability and longevity of the assets within these facilities.

In the case that mining and processing equipment is not commissioned or started properly, it can lead to catastrophic failure of assets within a short period of time. Standard operating procedures for facility assets, production operator inspection rounds to detect problems, along with an effective training programme for operations personnel, is expected to be developed to ensure the safe and efficient commissioning, startup, operation and shutdown for critical assets and other relevant mine and processing equipment. In addition, acceptance testing during commissioning is expected to be completed to verify installations have been constructed and operate as per the design.

Having experienced production operators directly involved in asset management strategy development and implementation greatly improves the asset management process and has been shown to increase the efficiency and longevity of assets. Trained production operators have knowledge to safely and effectively start and stop equipment, as well as operate equipment within design boundaries, which can enhance the mean run time between failure and lower overall operating costs. Production operators who are involved in routine equipment inspections can identify symptoms of pending asset failure and take the necessary actions (e.g. adjust or request maintenance repairs, or shut down equipment if catastrophic failure is imminent). Cross functional team assessment involving maintenance practitioners, maintenance engineers and operations personnel in FMEA processes for example, can identify probable failure modes that can be observed or measured during routine operation. Experienced production operators will play a role in identification of equipment distress signals that can be either observed or measured.

Operations personnel performing inspections of equipment are an important part of asset reliability. Production operators generally describes the operations personnel at a uranium mine or processing facility who are in direct contact with the equipment on a daily basis and are aware when the asset is running in a stable mode of operation. These personnel often have access to instrumentation based monitoring systems at their disposal to monitor equipment performance and health (e.g. pressure, temperature, vibration, electrical current draw). In addition, experienced and trained production operators can also rely on their senses (e.g. sight, hearing, smell, touch) to detect abnormal signals, usually associated with equipment in distress. As such, detailed inspection rounds ought to be developed for production operators to perform systematic and frequent inspection of equipment health and to detect problems.

As part of inspection rounds, trained production operators may also complete basic maintenance functions on equipment, when required, to reduce risk of premature equipment failure. When an operator is enabled to do basic maintenance functions, it allows the maintenance practitioners to focus on more complex maintenance tasks. A list of activities and basic maintenance tasks that may be completed by operations personnel to prevent equipment failure is provided below:

- Completing detailed work requests (i.e. equipment number, specific operational issues or equipment symptoms) in a timely manner after a distress signal is observed;
- Basic equipment lubrication;
- Bolt tightening;
- Equipment cleaning;
- Simple belt adjustments;

— Replacing filters or wear components (components designed to wear out and be replaced);
— Checking levels of lubricants, coolants;
— Collecting oil samples.

Production operator rounds for routine equipment inspections and basic maintenance need to be comprehensive and frequent enough to meet equipment inspection requirements, provide quantitative and qualitative record keeping as well as basic corrective actions or maintenance functions. Data from operator inspection rounds, including measurements, observations, corrective actions, or any other maintenance performed, are expected to be recorded on an inspection report and the data trends evaluated by the asset reliability engineer to monitor and track for equipment failure.

A production operator can provide an additional line of detection (e.g. visible vibration, noise, heat, etc.) if failures advance between predictive maintenance collection periods. The earlier an operator detects a symptom of failure (e.g. through instrumentation based monitoring equipment or through an operator inspection round), the more time the maintenance team has to plan and execute a repair. In some instances, simple maintenance actions can resolve the cause of distress and extend the time to failure or allow for replacement of the equipment in a planned manner to reduce downtime. Overall, production operator monitoring and inspections add an additional layer of protection against equipment failure as part of an effective asset management programme.

5.1.3. Reliability and defect elimination

As part of maintenance programme execution, the reliability team is responsible for defect elimination in asset management. Defects can be described as anything that erodes value, reduces production, compromises occupational health, safety, or the environment, or creates waste. The reliability team focuses on long term improvement which analyses historical data to find the major sources of downtime as well as reactive improvements for major equipment failures. Therefore, a focus on both long and short term issues is necessary for effective application of reliability engineering and defect elimination.

A defect elimination programme ensures that a systematic approach exists to identify and eliminate defects and ensures reliability resources are directed to have the greatest impact on the financial, environmental and safety performance of the facility. The defect elimination programme forms part of the continuous improvement element of the asset management framework.

Although a maintenance engineering team may take the lead with regard to implementing such a programme, effective defect elimination relies on inputs and accountability from a range of organizational stakeholders including production operators, maintainers, suppliers, warehousing staff, condition based monitoring technicians, amongst others.

The six steps of a defect elimination programme are outlined below and are further detailed herein:

(1) Identify and record defects;
(2) Assess and prioritize defects;
(3) Determine the root cause;

(4) Define actions;
(5) Implement actions;
(6) Measure outcomes.

5.1.3.1. Identify and record defects

Identification of defects is the first and most important step in the process. The identification of defects, over a period of time, comes from a range of sources including operations performance, work order feedback, operator/maintainer observations, online machine monitoring, data analytics programmes, amongst others, and captures both breakdown and nuisance events. Good practice would be to track all defects in a single register for traceability, where they can be categorized and further analysed.

During the ACA, productive units can be flagged for ongoing performance monitoring as part of the identification process. This requires an understanding of equipment requirements and performance such as:

— Availability;
— Operating regime;
— Built in redundancy;
— Performance (e.g. m^3/hour, temperature, tonnes);
— Operating cycle;
— Original equipment manufacturer design capacity and specifications.

The equipment performance can be graphically represented, such as in a Pareto chart, to display shortcomings as an indication of defects and the key performance indicators (e.g. equipment availability, maintenance costs, downtime events, etc.), and the trends evaluated over time. This process can be facilitated through the use of downtime analysis software which can automatically record and categorize downtime and faults to simplify the analysis process.

5.1.3.2. Assess and prioritize defects

Once all defects have been collated in a register, the impact of each defect and potential for reoccurrence can be reviewed. This assists with prioritizing the defects selected for further investigation. This step is reliant on collaborative discussion within a 'defect elimination meeting', which needs to involve personnel from maintenance, engineering, and production, as a minimum. This ensures that the defects to progress to investigation are prioritized and agreed upon across all operations teams. Involving a multi-disciplinary team in this process helps to identify equipment downtime issues, or other matters associated with downtime, providing a people-centric approach to defect elimination and accountability for asset health.

A common method to analyse failure data is a Pareto analysis. A Pareto analysis sorts the cumulative value of failures based on both frequency of similar events and the total impact, typically downtime, to the operation to allow simple identification of the largest contributors (e.g. frequent defects or downtime duration). The analysis is aligned to the Pareto Principle, also known as the 80/20 rule, where typically 20% of the issues will be causing 80% of the downtime. The most value can therefore be gained by focusing on the top 20% of failures. The remaining 80% of defect categories are then not selected for initial investigation but can be assessed and addressed once the highest frequency failures are resolved. The remaining defect categories may also be used for analysing data over a longer period.

The failure analysis observations are taken over a short period of time, typically a month, where the defects within that period are reviewed. Defects also need to be reviewed collectively over longer periods (i.e. 6 or 12 months) to capture lower consequence defects that may be happening more frequently. If the defect observation period is too short, smaller repeating downtime events may not draw attention, and may be missed for further investigation, despite them causing repeated problems. Over a longer period, these smaller repeat defects can be significantly more costly to the organization then a single large defect or failure. To help prevent this from occurring, a useful analysis is to include the total downtime experienced over a longer period, typically 12 months, to identify systematic defects. While major downtime events are typically chosen for investigation, it may also be necessary to select higher frequency downtime events. While these events may only contribute a small total downtime, they can be problematic as they cause disruption to the operations and frustration amongst operations and maintenance personnel.

The reliability and defect elimination process identifies which defects will be selected for further investigation. Root cause analysis (RCA) techniques are typically used for investigation, which is explored further in the following section. Investigation trigger criteria can establish when to perform an RCA, which may include for example, downtime loss, reduction in critical equipment performance or cost of the failure, amongst others.

Not all equipment requires the same level of reliability. For example, equipment that can be replaced easily, with low cost or that has a duty/standby arrangement with a run to failure maintenance plan. This equipment would therefore not require any RCA unless it was an unexpected or repeat failure. These reasons demonstrate why it is important to have a team of stakeholders to select and agree on the equipment defects to be investigated.

The defects selected for further investigation and analysis need to take in to account the resources available within the team to complete the investigation and implement actions. This resource availability needs to be taken into consideration during the selection process. If there are too many equipment defects selected for investigation, there will be no time for completing the actions from the investigations which is where the benefits of a defect elimination process are realized.

5.1.3.3. Determine the root cause

Once the defects have been selected for further analysis, individual investigations can be carried out. The most common types of RCA processes used, and their various properties, are outlined in Table 8.

The type of investigation analysis to be completed, along with the size of the investigation team and special knowledge of individuals invited to participate in a root cause analysis needs to be defined by the defect elimination programme, typically with input from the organization’s senior management team in making these decisions. Smaller defects or less complex investigations may be able to be assigned to a single person or a small work group to investigate.

A defect investigation generally requires analysis of historical records and performance measures including the following:

— Production and availability targets;
— List of failure modes for equipment;
— Work order history;
— Cost;
— Production reports;
— Existing equipment strategy;
— Duty/standby arrangements;
— Time to deliver spares;
— Cost of spares;

All investigations are to be completed in a timely manner, ideally within 60 to 90 days, to ensure completion of preventive and/or mitigatory actions, but also to reduce the likelihood of defect reoccurrence before those actions can be implemented.

TABLE 8. TYPE OF ROOT CAUSE ANALYSIS PROCESSES

Root cause and problem solving tools	Brain-storming	5 Whys	Fish Bone	Cause and Effect Tree	TapRoot	XYMatrix	FMEA	Proact / Apollo
Efficiency at finding multiple critical factors	Low	Low	Low	High	Medium	Medium	High	High
Typical time to run	<30mins	<30mins	<30mins	1–2 hours	1–3 days	1–2 hours	<1 day	<1 day
Structured approach	No	No	Yes	Yes	Yes	Yes	Yes	Yes
Encourages divergence (multiple causes)	Low	Low	Medium	High	Medium	Low	High	High
Converging – Prioritizing the causes	Weak	Weak	Weak	Good	Medium	Medium	Good	Good
Data driven or observation	Observation	Observation	Observation	Data	Observation	Data	Data	Observation
Focus on severity, occurrence and detection ability	No	No	No	No	No	No	Yes	No
Visual communication tool	Weak	Medium	Good	Good	Medium	Good	Good	Medium
Living document, can use to manage implementation	No	No	No	No	No	No	Yes	Yes
Can use proprietary software	Yes	Yes	Yes	Yes	Yes	Yes	Yes	Yes
Facilitator competency	Low	Low	Low	Medium	High	Medium	Medium	High

5.1.3.4. *Define actions*

Once a defect investigation has been completed, a series of actions need to be developed that would prevent a defect reoccurrence or mitigate the consequences of failure. There is rarely a single factor which causes a defect. More commonly there would be up to 4–8 contributing factors, with one factor in particular which would have prevented the defect from occurring. The identified contributing factors need to be reviewed and a decision made regarding what actions are going to be implemented. A process of engagement ensures all relevant stakeholders are informed about the investigation outcomes and have input into action development, ultimately taking accountability for the improvement activities agreed upon. The time taken to implement actions needs to be weighed up against the value gained. Some actions will be very costly and require significant resource for a relatively small benefit. It is good industry practice to share defect investigations amongst other relevant business units within the organization, such that opportunities to replicate improvements can be utilized.

5.1.3.5. *Implement actions*

Once defect elimination actions have been decided on, they need to be assigned and tracked using an action tracking system that is endorsed by the organization. Action tracking needs to ensure that all actions that were selected for implementation are completed. By not tracking actions from the RCA there is a chance that the investigation is completed to a high standard, but the actions do not actually get implemented, which increases the likelihood of defect reoccurrence. To ensure all actions are completed, and follow correct processes, actions may be integrated into the work management systems or CMMS. All tasks associated with the completion of action need to be documented, for example MOC forms, updating maintenance documents, updating equipment drawings, updating document registers, updating key performance indicators and measurement outcomes, etc.

5.1.3.6. *Measurement outcomes*

Defects that have been investigated are expected to be revisited to ensure the actions implemented have sufficiently reduced reoccurrence and to decide if further action needs to be taken. Measuring the outcomes of defect elimination actions needs to be done at a defined period, usually within six months after actions have been implemented. A simple way to measure outcomes is to track the defect over a period, before and after the actions were implemented, to check that repeat failures or other defect issues have been resolved. The failure frequency and mechanisms ought to be considered when measuring outcomes, as insufficient time may have passed for the issue to develop. It may be difficult to measure the effectiveness of actions for low likelihood larger failures scenarios, however it is still worthwhile completing this exercise.

5.2. WORK MANAGEMENT

Work management is necessary to ensure that competent persons are correctly executing maintenance work on assets. Work management is a systematic process in any organization, as all asset management strategies that support the operations are ultimately deployed through a work management process using the CMMS.

Work management is an integral part of an organization's maintenance framework. Work management is at the centre of maintenance management and focusses on executing work.

Work management can be systematically and chronologically arranged into six main steps. The six steps of work management are defined as:

(1) Identify;
(2) Planning;
(3) Schedule;
(4) Execution;
(5) Record;
(6) Analysis.

5.2.1. Identify

The work management process begins with work identification which clearly and succinctly identifies maintenance work that needs to be performed. During this step, the maintenance work that is required to maintain the health of assets are identified either from operator observation, primary maintenance or additional maintenance. For example, identification may arise during maintenance work based on asset tactics, work arising from inspections, from failures or condition monitoring, or safety and continuous improvement activities. This work is initially submitted as a notification for further approval and processing in the work management cycle.

Primary maintenance work is based on the asset maintenance tactics developed through the FMEA process, as covered in Section 4.4. These tactics are therefore in essence pre-approved work tasks for specific assets. These tasks will already have well defined plans that includes the steps and time required to complete the work, the parts and tools required, any specific safety risks and mitigating strategies (e.g. defined PPE, working at heights tie off requirements or special rigging and hoisting for lifting equipment). These maintenance tactics and work packages are loaded in the CMMS and have a predetermined interval at which they are triggered. The triggering of primary maintenance work can be automized based on equipment usage, which is either time based or based on cumulative production throughput. Where automatic triggering is not available due to data, tactics are manually triggered.

Additional maintenance work can come from many sources. It may come from the maintenance department as a result of inspection activities, breakdown or condition based monitoring. The operations department is another area from where additional maintenance work is generated. The work identified from operations may be based on operating conditions such as deteriorating equipment performance producing poor product quality. To assist the planning and scheduling of additional maintenance work, it is necessary to indicate the timeliness or priority to which the work needs to be completed. For example, if a repair is urgent in nature, then it may break into the maintenance schedule and may be completed without following the planning processes. In practice, maintenance strategies ought to allocate less than 15% of the total maintenance work as this type of reactive work. Additional maintenance work may be generated by production areas, but also areas related to environmental, occupational health and safety matters. Additional maintenance work may also be generated as part of an engineering project or by equipment improvement suggestions [30].

During this early stage of the work management process, the work required is to be identified with as much detail as possible to allow efficient planning of the work. The level of detail would include as a minimum the scope of the work, the personnel and skills required, the duration, equipment and materials required to perform the work.

5.2.2. Planning

The planning step in work management is the process of developing the detailed scope of work that defines the requirements necessary to execute the work.

This process begins with pulling together the list of submitted notifications. The task steps to complete the work are put together in a logical sequence.

These task steps are then consolidated into logical work packages, incorporating all necessary work procedures, labour and skills required to execute the work, together with parts, materials, tooling, equipment for each task and any specific safety work requirements and relevant risk mitigations. Good maintenance programmes have task steps created three to four weeks in advance of the work being executed and are identified in the CMMS as 'ready to schedule'.

A plan for a specific task ideally needs to be done only once. The plan is stored in the CMMS and reviewed from time to time based on the needs of the operation. In essence, a good work order contains the following information:

- Special safety instructions including isolation lists and lock out/tag out instructions specific for this maintenance work;
- Detailed instructions including relevant equipment drawings;
- Sequencing tasks;
- Planned work duration;
- Labour and skills required;
- Equipment isolation requirements (e.g. from energy sources such as electrical, mechanical, gravitations, etc.);
- Materials and spares required;
- Permits required prior to executing the work (e.g. hot work, confined space or radiation work permits);
- Initial cost estimate;
- Forms for recording measurements.

5.2.3. Schedule

A maintenance work schedule summarizes all resources that are required within a specified time to execute maintenance work. In a uranium mine and processing facility, maintenance work schedules are usually prepared on a weekly basis. Developing the weekly schedule is the responsibility of the maintenance scheduler. An accurate work schedule requires that the planning process indicates in detail all required resources (labour and materials) and duration for the tasks. Additionally, the schedule relies on accurate and timely information and feedback from maintenance planners, maintenance tradespersons and operations personnel. It is imperative that the maintenance schedule be accurate, because it acts not only as a mechanism to schedule work, but it also performs as a communication tool.

As equipment and assets are the subject of maintenance, the scheduling of maintenance tasks needs to consider the organization's requirements, specifically the production plan and product sales projections. A review of the maintenance work schedule against the organization's requirements is required to understand how equipment downtime, required for maintenance, may impact the organization's profitability [30].

To ensure effective use of the maintenance tradespersons' time, the equipment to be maintained is isolated, cleaned, locked out by operations personnel and is ready for the maintenance tradespersons to begin work at a time predetermined in the weekly maintenance schedule. This requires strong coordination between the maintenance and operations teams, which is facilitated by the weekly maintenance schedule.

The work schedule is typically developed and endorsed by relevant organizational stakeholders (e.g. representatives from maintenance, operations, safety, radiation protection departments) 5–7 days before the work is executed. The mechanism of communication ensures that the proposed maintenance plans are endorsed or modified where required.

Finally, if the maintenance task is non-routine and has an elevated safety, radiation or environmental risk, some form of risk mitigation process (e.g. field level risk assessment or job hazard analysis) is to be completed with all relevant stakeholders in advance of the work being executed. This exercise may take several hours and therefore is also included in the weekly schedule. This activity is generally scheduled the week prior to that in which the task is scheduled to be completed.

When developing the schedule, the maintenance scheduler needs to consider some of the common constraints in uranium mines and processing plants:

- Spares availability: The scheduler works with warehouse personnel to ensure the parts required to complete the maintenance tasks are available and prepared prior to the maintenance task being scheduled.
- Availability of trades: Ensure the schedule is not overloaded relative to number of trades hours available.
- Availability of equipment: Ensure the equipment available such as cranes, scaffolding, vacuum trucks, welding equipment, etc., are in working order and not double booked.
- Production plan: Ensure that operations are able to hand over the equipment to the maintenance team for a predetermined amount of time without risk to production, safety or environmental impact.
- Radiation protection: The scheduler works with radiation protection experts to ensure that required radiation surveys are completed and monitoring equipment is available. Ensures that arrangements are in place to optimize radiation dose, such as scheduling the rosters of maintenance tradespersons to limit exposure times.

The key output from the process is a schedule that is:

- Loaded with labour with an allowance for reactive work (e.g. 10–15% allowance for urgent or unplanned maintenance work);
- Developed in accordance with the operations plan;
- Optimized in the allocation of resources including labour, materials and equipment;
- Budgeted, or otherwise financed with approval;
- Endorsed by all key stakeholders.

Scheduling can be summarized as the step in the work management process whereby maintenance activities can optimally be executed.

5.2.4. Execution

The next stage of work management includes those steps that are directly associated with the execution of maintenance work. Commencing with the assignment of work to designated maintenance teams via work orders and/or work packages and in keeping with the requirements of the weekly maintenance schedule, but also any immediate issues with availability of labour, materials and equipment. Prior to commencing work execution, risk assessments are conducted, permits are authorized and appropriate materials, tools and equipment obtained. In addition, all equipment components and parts required to complete the work task ought to be staged at the work face by the warehousing staff.

The maintenance task is executed in accordance with the detailed work order and progress is monitored throughout. In the event that additional work tasks are identified, these will be undertaken as a revision of the existing open work order (along with a revised risk assessment, permits, etc., as required) or alternatively as a separate work order with a new notification. The approach ought to be decided based on discussions between maintenance and operations senior management.

The execution of tasks requires cleanup of the work area after the maintenance is completed. A detailed work order includes cleanup tasks, which are typically assigned to maintenance trades personnel as responsible for cleanup of the work area and tools.

As part of task completion, maintenance trades workers and operations staff are expected to conduct a commissioning step of the maintained equipment to ensure it is ready for duty. This step is performed as final authorization of the work order by the maintenance and operations personnel.

5.2.5. Record

All relevant information and data associated with maintenance tasks, needs to be appropriately captured on the work order for record keeping and for further analysis. This information includes documentation, technical records, quantitative data, etc., as well as qualitative feedback from tradespersons. This step is important for cost and performance measurement, but also important for ongoing improvements in reliability, maintainability and overall asset availability. Documenting of the relevant information and data is completed by whomever executes the work (i.e. the maintainer).

Record keeping is required to ensure that rotable and repairable items are set aside for collection, and any extraneous parts that remain following completion of the maintenance work are properly returned to the warehouse.

Record keeping is expected to also capture all post-execution tasks and administration of maintenance reports including confirmation of hours and the collection of maintenance history. Records of all resources used, delays experienced and additional work required is captured in CMMS during this step. This history ought to be reviewed by maintenance senior management personnel, and edited, if required, prior to entry into the CMMS. Entry into the CMMS may be completed by maintenance tradespersons, the maintenance planner or a maintenance clerk.

5.2.6. Analysis

The analysis of work orders facilitates the continuous improvement of maintenance work management. It involves:

— Analysing and improving the work management process itself, including how the process is effectively deployed and complied with;
— Analysing and improving overall asset performance, including asset reliability.

Specifically, the analysis process includes routinely reviewing key performance indicators, daily and weekly review meetings and managing and completing improvement action plans. The analysis stage also determines whether the maintenance tactics and operations performance are producing the required reliability.

5.3. PERFORMANCE MANAGEMENT AND CONTINUOUS IMPROVEMENT

Asset management requires performance management and continuous improvement efforts, similar to other organization functions, in order to optimize strategy and realize best value.

A basic continuous improvement process is shown in Fig. 8 'Plan – Do – Check – Act', which finds its basis in the Kaizen philosophy of continuous improvement, which can be adopted by a broad range of organizations and applied to all functions [28]. Within uranium mines and processing facilities, the 'Plan – Do – Check – Act' methodology is often applied in areas of occupational health, safety and environmental management, but this also has application for asset management for the purpose of achieving orderly and gradual continuous improvement.

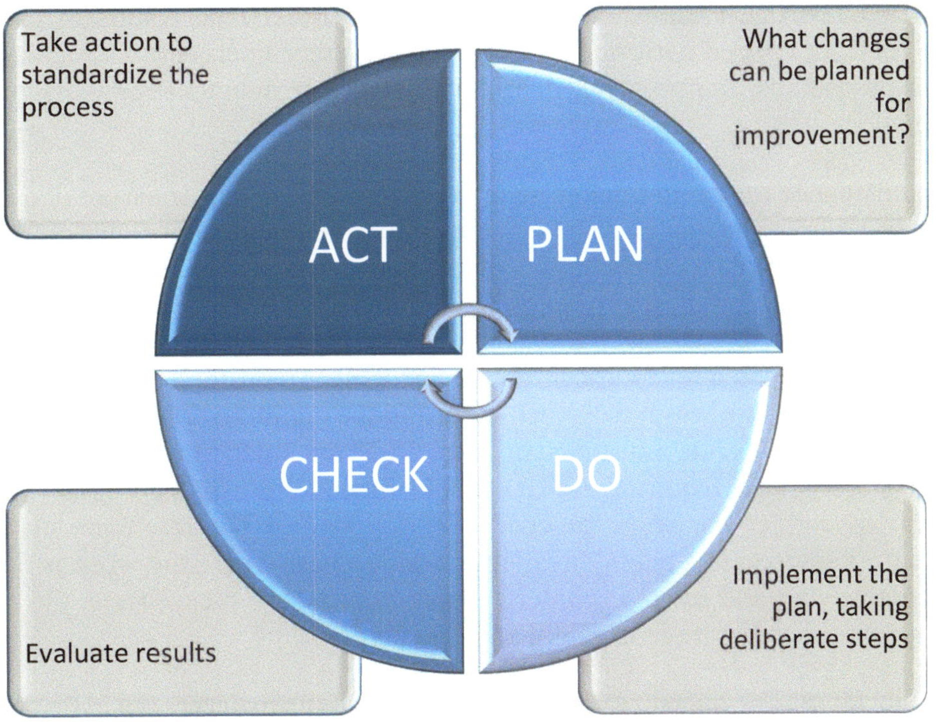

FIG. 8. The continuous improvement process.

The tools of 'lean production' can also be applied in order to improve the efficiency of asset management planning, scheduling and execution. Lean production refers to an operational approach that eliminates non-value-added activities through the entire production chain. For instance, eliminating defects requiring rework, removing unnecessary processing steps, eliminating waiting times, excess inventory, amongst others [28]. An example of how the principles of lean production can be used in maintenance tasks include coordinating the set-up and location of the maintainer's tools so that they are readily available in the work area if there is a need for flexible tool selection or minor adjustments to perform the task. Another example would be to have spare parts strategically located so that equipment can be rebuilt in a very efficient manner, with the work procedures available, and tools packaged along with the parts, thereby reducing the total time required to repair the equipment.

Statistical analysis can be utilized to track performance of maintenance processes and to identify deviances from a predetermined standard. Data can be useful to understand functional failure conditions, which may not be readily identified using standard inspection techniques. Statistical analysis methods require large data sets, which would typically come from the digital process control system, and then analysed by a specialist such as a reliability engineer. The outcome of the process includes improvement opportunities and changes to the asset management system to correct the deviance.

The system by which maintenance work is identified, planned, scheduled, and executed is very often tracked and analysed to optimize the effectiveness of the maintenance organization. For example, in complex industrial environments such as uranium mining and processing facilities, there may be many items that require maintenance attention. Not all items can be addressed at the same time. As such the identified maintenance work needs to be reviewed, approved, and ranked according to priority, before proceeding to the planning stage of work management. This process would be categorized as a sustainability audit of the maintenance work management system and can be measured against the operation's work management predetermined limits (i.e. established performance measures). The results of this audit may indicate the need for revised performance measures, or other interventions such as changes to maintenance procedures, training, etc., to achieve effective maintenance work processes for asset management.

A typical maintenance metric to track is schedule compliance. This measures how effectively the maintenance work schedule is executed. Typically, the maintenance work schedule would be set the week before planned execution of the tasks. During the period of execution, the schedule is to be followed, as best as possible, to execute the work in a coordinated and efficient manner. If there are interferences with the schedule, then these need to be tracked and identified, so that they can be corrected for future work management. An example of this could be that the work procedure for the task was not correct, requiring the maintenance tradesperson to identify correct reference materials (i.e. industry standards, OEM documents), which may not be readily available in the workplace, requiring the task to take much longer than necessary. A solution to this could be to correct the work procedure or to provide the reference material to the maintenance tradesperson in a readily available location near where the work is occurring. By continually identifying and removing interferences to schedule compliance, the execution of maintenance work becomes more efficient and effective.

5.3.1. Key performance measures for work management process

Performance management of assets relies on key performance measures, or also known as key performance indicators. Table 9 provides a series of performance metrics that can be adopted to measure the effectiveness of the work management process in uranium mines and processing facilities. Each of these performance metrics have interdependencies and the value of each is to be taken into context holistically, within the entire asset management strategy and maintenance programme. Focusing only on one specific performance measure may have unexpected results and a detrimental impact on the other aspects of facility maintenance or operation. To achieve effective continuous improvement, trained and experienced maintenance professionals are needed to understand the relationships between performance measures and the organization requirements, to deliver on the production plan and product sales projections.

TABLE 9. KEY PERFORMANCE MEASURES FOR WORK MANAGEMENT

Metric	Metric definition
Backlog	The amount of previously scheduled maintenance work that has not started. Enables the scheduler to identify which maintenance activities need to be rescheduled or closed/cancelled
Forward log	A measure of maintenance workload coming up in the future, including work that is being deferred to subsequent weeks. Enables the scheduler to identify the need for additional resources, or to reschedule to suit available capacity
Outstanding notification (%)	Notifications that are awaiting review and approval to proceed to planning and scheduling
Outstanding work	Outstanding work is a measure of all work identified, that is every task within the work management system. It may be considered a profile of work constituting backlog and forward log (i.e. both past and future work respectively)
Overdue regulatory maintenance (statutory compliance)	A measure of compliance to regulatory maintenance, in accordance with statutory requirements
Primary maintenance schedule compliance (%)	A measure of completion of primary maintenance work within the scheduling period, and compliance to preventive maintenance
Resource utilization (%)	A measure that indicates the percentage of total trades capacity available compared to the total hours that has been captured on all completed work orders. This is an important measure to determine whether the full capacity of the trades is being effectively utilized
Schedule compliance (%)	A measure that monitors the percentage of scheduled work operations completed, relative to all work operations completed in the period. It is an indicator of the level of control that exists over the work management process
Rework (%)	Work that is carried out on the same maintainable item within 24 hours of the original service/repair or within 24 hours after it was put back in service following scheduled maintenance. This is a measure of the quality workmanship and/or compliance to procedural specifications and standards
Scheduled work percentage (%)	A measure that monitors the percentage of scheduled work hours completed to all work hours completed in the period. It is an indicator of the level of control that exists over the work management process and how well the operation protects the schedule

The Society for Maintenance and Reliability Professionals has published a document titled SMRP Best Practices that provides a standard set of performance metrics that can be used to track how well assets are functioning and how well the asset management system is performing [31], which can be applied for use in uranium mines and processing facilities.

5.3.2. Time model

To ensure consistent definitions are used across the operations for which the performance measures are based, a 'time model' ought to be developed which defines the calculation for the time usage model. There are many ways to produce a time model, which necessitates the development of a standardized model within an organization. An example of a time model is outlined in Fig. 9.

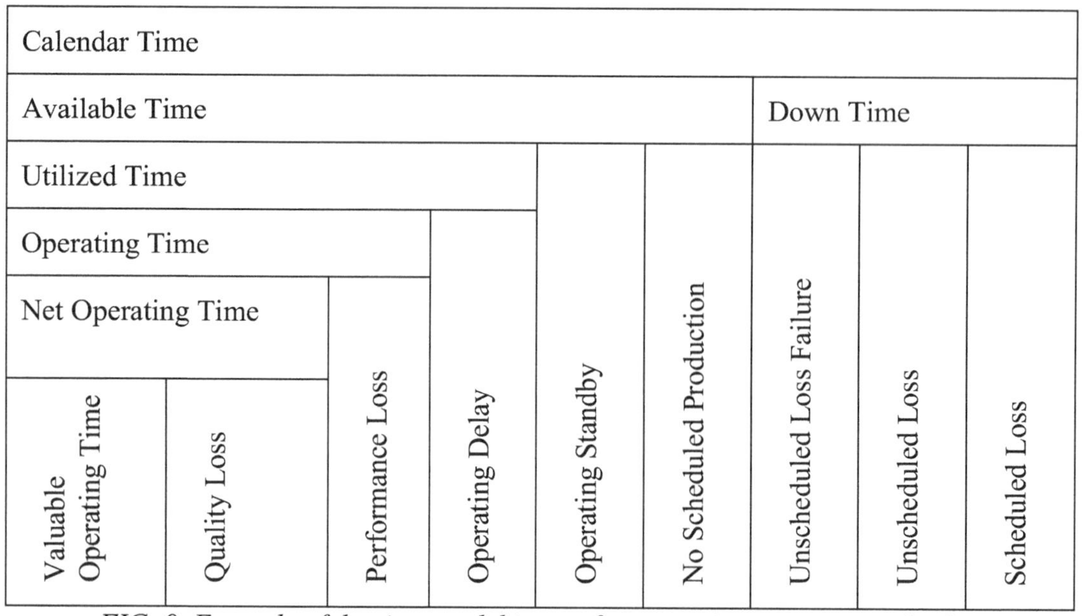

FIG. 9. Example of the time model in use for a uranium processing facility.

When undertaking performance benchmarking against other operations, organizations or industries, it is necessary to ensure that the time model equations are the same, or otherwise make adjustments to them, to ensure accurate comparisons are made.

Standardized definitions for time models can be found in Ref. [31]. Another resource for time model information is Ref. [32].

5.4. REPAIRS AND MODIFICATIONS

Industry good practice for repairs and modifications performed on assets ensures that such activities conform to the relevant engineering or equipment standards. Repairs and modifications need to be appropriately tested, recorded and inspected. Repairs and modifications are performed by competent personnel only and inspected by a competent engineer.

Major repairs and modifications of registered equipment need to be captured in the organization's registered equipment database. Prior to return to service, these repairs and modifications need to be recorded, and copies maintained within the documentation files for all registered equipment. Repairs and modifications to registered equipment are also expected to be appropriately tested and recorded. In some cases, repairs or modification may require a certified inspection to be completed by a competent engineer before the asset can return to service.

5.5. KNOWLEDGE MANAGEMENT

Knowledge management can be described as the process of organizing the knowledge and information of an organization. Each operation may consider having documents and drawing management systems in place to ensure that these important pieces of information are controlled, properly stored, maintained, accessible and kept current, as well as maintaining historic information. These documents typically include:

— Mine and processing facility infrastructure design drawings;
— Equipment specifications (including OEM documentation);
— Operating procedures and manuals;
— Records of assessment (e.g. risk assessment, ACAs, tactics records);
— Work procedures;
— Equipment warranties and guarantee correspondences;
— Maintenance manuals;
— Test certificates;
— Reports;
— Piping and instrumentation diagrams;
— Other drawings, including construction and as built drawings.

Physical drawings and files are to be kept in a database or document vault. It is good practice in asset management, to consider keeping hard copy file records for the entire lifetime of the operation, and even past the operational lifetime of an asset itself.

The CMMS and asset design software for life cycle costing are effective tools for budgeting and performance management purposes. It is important to ensure accuracy and efficiency of access to electronic data. Examples of master data which is typically loaded in a CMMS includes all the files and associated information such as bills for materials, drawings, manuals and work procedures. It is beneficial for operations to have master data readily available for designers, maintainers and engineers.

The organization is expected to develop guidance for information management and information security management, by which documents, in hard and electronic formats, will be maintained. The accountability and resources for the control of these documents also needs to be defined. In complex organizations like uranium mines and processing facilities, this can be a significant effort, requiring accountable resources within each operations department as well as the operation overall. The responsibilities for document control includes the following:

— The important documents are to be kept secure either in electronic format, hard copy or on magnetic storage media. Aspects related to security of this information such as fire protection, cyber security and misplacement need to be carefully considered.
— New documents added to the system are to be recorded and indexes kept up to date.
— The versions of documents and drawings are to be tracked when original documents are removed from storage or filing systems.

Auditing the effectiveness of the knowledge management system on an annual basis is an effective mechanism to determine the level of accuracy of an operation's drawings and documents.

5.6. TROUBLESHOOTING

Troubleshooting guides for critical equipment may become available through official channels (i.e. manufacturer releases, statutory gazettes) or developed internally by the organization through knowledge management and feedback from maintenance and operations personnel. Troubleshooting guides need to be readily available to maintenance personnel (including professionals and trades workers). Consideration may be given to electronic platforms, for example an operation's computer, intranet or any other accessible reference system across the operations (e.g. CMMS). Alternatively, hardcopy guides may be required in remote locations where access to electronic platforms is limited. This information is to be continually updated and developed (including pictures where appropriate) to ensure that information regarding troubleshooting, operational knowledge and lessons learned are utilized to ultimately ensure that unnecessary downtime is minimized.

6. DEGRADATION MECHANISMS AND CONDITION MONITORING

Fundamental to realizing the best value of an asset, is maximizing its useful life for the lowest reasonable cost, while still maintaining the quality of its intended function. For example, the function of a process tank (or vessel) is to hold a certain volume of fluid. If that tank were to develop a leak, then it is no longer serving its intended function. Furthermore, if that tank were to become partly full of unwanted solid material, then it is still holding fluid, however not to the full volume intended, reducing its usefulness. In this simple example, it would be preferable to be able to identify indicators of loss of function or usefulness in the lead up to the undesirable event, before the negative effects are realized, and to take affirmative action to correct. Degradation mechanisms and equipment monitoring methods used to proactively identify potential equipment failure or loss of function can be relatively simple, but they may also be very complex, requiring specialized training and equipment. This section provides detail on a variety of general techniques and technologies to monitor the physical condition of assets (see Section 6.1) as well as more specific methods that are typically used for the common degradation mechanisms observed in uranium mines and processing facilities, including fatigue, corrosion, stress corrosion cracking, flow accelerated corrosion, abrasion and thermal ageing, amongst others (see Sections 6.2–6.14).

6.1. CONDITION BASED MONITORING TECHNIQUES

Condition based monitoring provides an understanding of the asset health of mining and hydrometallurgical processing equipment. Through understanding the condition of equipment and having capability to detect developing faults, maintenance strategies can be created or adjusted to prevent failure and increase the overall availability and lifespan of the equipment. This also has the benefit of reduced life cycle costs of the asset. Condition based monitoring can be classified as preventative, predictive or proactive and typically includes analysis of oils and lubricants (tribology), vibration (vibration analysis), temperature (thermography), electrical current (insulation resistance testing) and corrosion or cracking (non-destructive testing). These techniques may be used concurrently to gather a holistic understanding of the condition of equipment.

Due to the cost associated with performing condition based monitoring, it is typically reserved for equipment associated with critical assets. The decision making process for determining which equipment requires condition based monitoring is part of the ACA (see Section 4.2), with the techniques and frequency of monitoring being developed during the tactics development stage of asset management (see Section 4.3). Primary predictive monitoring techniques are detailed within the sections below.

6.1.1. Oils analysis (tribology)

All rotating equipment within a uranium mine and processing plant relies on lubrication to protect surfaces that are moving against each other to reduce friction and wear. The type and quality of the lubrication has a significant effect on the ability to reduce friction or wear and directly affects the operational life of the equipment. There are many types of lubrication available, and the correct lubricant needs to be selected for each application.

Contaminants within the lubricant will reduce the effectiveness of the lubrication and can result in significant damage to equipment. The most common contaminants are water and dust, however there are many other contaminants within oils which reduce its effectiveness. Periodic analysis of the contaminants within the lubricant can indicate if the lubricant is becoming contaminated from external sources or if the equipment it is protecting is wearing, as well as a range of other issues. The results of oil analysis (tribology) can be used to determine how and when the oil needs to be changed, and to also ascertain the health of the equipment. This information will assist with planning equipment maintenance or replacement. Oil analysis is the most sensitive condition monitoring technique and will, in most cases, detect potential issues before vibration analysis or thermography techniques can detect issues. To ensure an effective oil analysis programme, an in depth understanding of the types of contaminants that would be produced within the equipment is needed, so that the correct testing and analysis techniques can be performed to detect those contaminants.

Oil samples are expected to be analysed by a trained condition monitoring technician. Oils analysis is typically only performed on critical equipment at prescribed frequencies (e.g. monthly or quarterly). The samples are usually sent to an external qualified testing facility to perform specialized analysis. Many lubrication suppliers provide oil testing services as part of their contract to supply lubricants to the mine and processing facility. In some cases, the supplier may perform oil analysis as part of the overall contract arrangement. The operation's asset management strategy needs to be appropriately resourced to ensure that the oil analysis includes the range of testing required to perform an effective evaluation of the equipment. Following testing of the oils, the reliability engineer or condition monitoring technicians are required to make risk based decisions to maintain, replace or continue monitoring the equipment.

Further guidance on oil analysis, with application to uranium mining and processing equipment, can be found in Refs [33, 34] (see Appendix VIII).

6.1.2. Vibration analysis

Vibration analysis is the measurement of equipment vibrational frequency and amplitude patterns to detect possible failure modes. Vibration measurements are taken on specific locations on equipment, generally using a small flat sensor pad made for vibration monitoring of each axis of the area being measured. The signal is received by a vibration meter which is then either analysed by a trained technician in situ or loaded into a computer for further analysis using vibration analysis software. Due to the cyclical nature of rotating components, each component will have a frequency at which is vibrates. In general, larger amplitude measurements at particular frequencies signal a potential failure pattern for that component.

Vibration analysis is extremely sensitive and will detect minor issues in equipment prior to most other detectible issues, for example increases in temperature, noise, or smoke. Detection of abnormal vibrations can allow preventive maintenance to be taken to prevent further damage. By analysing trends in the size and amplitude of the vibrations, the severity of the issue over time can be used to determine an approximate component replacement date before it becomes an issue. Vibration analysis can be used to detect issues such as shaft misalignment following installation, lubrication issues, bent shafts, unbalance, resonance, loose parts, bearing damage, worn or damaged gears, amongst others. Like all condition based monitoring, vibration analysis is generally reserved for critical equipment due to the resources required to maintain, collect and analyse data. Modern equipment is more readily available with vibration sensors pre-installed, due to vibration sensors reducing in cost. The installation of permanent sensors is also useful in difficult to access locations where manual readings cannot be taken, for instance behind equipment guarding.

Vibration monitoring is typically completed by in house condition monitoring technicians who are trained and certified in vibration analysis. These technicians take the readings, analyse the data and raise work notifications where necessary to further investigate or replace equipment.

Further guidance on machine vibration analysis can be found in Ref. [35] and in its related publications in the series (see Appendix VIII).

6.1.3. Insulation resistance testing

The insulation resistance of a motor or electrical components degrades over time with the ingress of moisture and dust, extreme temperatures, mechanical damage or electrical stresses. By ensuring the insulation resistance of equipment is maintained, the potential for breakdown or electric shock can be minimized. Any measured reduction in insulation resistance are to be reviewed, and a strategy implemented to replace the equipment. The equipment used to conduct insulation resistance testing is called a megohmmeter.

Insulation resistance testing is usually required as part of a uranium processing facility's insurance policy, with particular application within the solvent extraction area of the processing facility to ensure that no sparks are able to start an electrical fire.

Following a fire or if electrical equipment has been dropped or experienced any other stresses, an insulation resistance test ought to be completed to ensure the integrity of the equipment.

Insulation resistance testing is expected to be completed by qualified electricians and is typically only completed on critical equipment.

Further guidance on insulation resistance testing can be found in Ref. [36] and its related publications in the series (see Appendix VIII).

6.1.4. Thermography

Thermography, or infrared imaging, is used to reveal abnormal temperature variations in both mechanical and electrical equipment. A thermal imaging camera is used to instantly view temperature variations on equipment. For electrical equipment, a 'hot spot' will identify the location of an overheating component prior to it failing. This can prevent unplanned downtime of electrical equipment which can be difficult to determine. Unlike mechanical equipment, where failure can be detected through a variety of condition based monitoring methods, electrical equipment will often only overheat.

Thermal imaging allows the condition monitoring technician to detect issues from a distance without having to come into contact with potentially hazardous components. Thermography often requires the thermographer to open electrical cabinets which is only be completed by trained and qualified electricians, using specialized equipment and PPE.

Reference [37] requires thermal imaging to be completed by a qualified thermographer.

Within uranium processing facilities, thermography has some unique applications including:

- Grinding mills: Detecting areas on the shell of the grinding mill where uranium ore may be in contact with the shell signalling an issue with the liners, bolts or rubber lining behind the mill liners.
- Tanks: By using thermography over the entire shell of a tank, the 'bed' level within the tank can be detected, indicating if an increased layer of (process) solids have settled at the bottom of the tank. This can signal a process upset or poor agitation due to damaged or broken agitator blades.

Hot spots on the shell of a tank can signal failure of the rubber lining of a mild steel tank where product is coming in to contact with the shell. Rubber lining failure is of particular concern in leaching, CCD and neutralization tanks due to the nature of uranium process fluids and slurries. Early identification of a liner failure may prevent holes or catastrophic failure of a tank. It is noteworthy that thermography of tanks can be difficult because of varying external temperatures of the tank, particularly during the day when there is direct sunlight on the tank surfaces, causing heating at different rates. The application of thermography is therefore best performed at night when the external shell temperature is more consistent. Due to the proximity of tanks, it can also be difficult to get a thermographic image of the areas within the middle of a group of tanks.

Further guidance on thermography can be found in Ref. [37], and its related publications in the series (see Appendix VIII).

6.1.5. Non-destructive testing

Non-destructive testing (NDT) is any testing that can be performed without detrimentally affecting the equipment. This testing is often used on operating equipment as it allows equipment to be tested while online or with minimal downtime. NDT is used to detect faults within materials to allow an assessment to be made if the equipment can remain in service, requires repair, if further monitoring is required or if it needs to be taken out of service.

Common NDT inspection methods are outlined in Table 10 and the advantages and limitations of each of the methods is described in more detail in the following subsections.

TABLE 10. COMMON NON-DESTRUCTIVE TESTING INSPECTION METHODS

NDT inspection method	Common industry abbreviation	Description of application
Phased Array	PA	Advanced ultrasonic technique for detecting cracks both internally and externally
Magnetic Particle	MPI	Standard technique for detecting cracks. Needs to be fluorescent based for stress corrosion cracks
Visual	V	Used in all cases (after cleaning) as an initial inspection to identify macro deterioration in structures
Ultrasonic Thickness	UT	The determination of remaining wall thickness is easily determined with ultrasonic techniques. Crawlers can be used where access is difficult (refer to rapid motion scanner)
Rapid Motion Scanner	RMS	RMS is a high speed, high accuracy remote access ultrasonic corrosion mapping system designed to capture thickness on a broad scale
Eddy Current Field Measurement		An electromagnetic technique for detection and sizing of surface breaking cracks
Saturated Low Frequency Eddy Current	SLOFEC™	Specialized technique for corrosion mapping for under floor corrosion using eddy current along with magnetic field
Dye Penetrant	DPI	This method is used to locate the surface breaking defects in all ferrous and non-ferrous material
Acoustic Emission		This method is used to detect any cracking/leaks. More often used for pressure vessels
Magnetic Flux Leakage	MFL	Used to scan the floor of a tank through the lining. Cannot scan through thick liners. Similar to SLOFEC™ however this method can determine if corrosion is topside or underside
Radiography		Portable X ray equipment to see through the profile of the steel, often used for weld testing

When planning for maintenance activities, consideration needs to be given to the radiation protection requirements when using NDT equipment using ionizing radiation.

6.1.5.1. Phased array inspection

Phased array (PA) inspection is a type of ultrasonic testing method that uses multiple transducers that are pulsed with various delays (hence phased) respective to each other to create a 'wave', which can be directed at various angles and swept over areas for scanning.

Advantages:

— Higher accuracy when compared to standard UT testing;
— Provides immediate feedback to technician;
— Detailed colour images can be produced showing location, size and shape of defects.

Limitations:

— High equipment costs;
— Further training required, in addition to UT testing.

Standards have been developed that provide guidance on performing PA inspections. Further guidance on PA inspections can be found in Refs [38, 39] (see Appendix VIII).

6.1.5.2. *Magnetic particle inspection*

Magnetic particle inspection (MPI) works by detecting irregularities through the application of a magnetic field acting on a piece of ferrous magnetic material. In essence, an electromagnetic force is applied to the test area. The flux passing through the test area will be disrupted by cracks and voids which force the flux outside of the test area. A fluid containing electromagnetic particles is placed on the test area where the particles are attracted to the flux at the surface. The fluid also contains pigments which fluoresce under ultraviolet light making them easier to detect.

Advantages:

- — Simple surface preparation;
- — Rapid inspection for large parts;
- — Surface preparation is less important;
- — Low cost for testing equipment;
- — Method allows technician to visibly see the defect on the test piece in real time.

Limitations:

- — Only works for materials that can be magnetized (i.e. ferromagnetic materials);
- — Large test pieces require larger currents;
- — Ability to detect defects is sensitive to the alignment of the magnet with the defect;
- — Surface required to be uncoated, relatively smooth and demagnetized prior to testing.

Standards have been developed that provide guidance for performing MPI. Further guidance on MPI can be found in Ref. [40], and in Ref. [41], and its related publications in the series (see Appendix VIII).

6.1.5.3. *Visual inspection*

Visual inspection (VI) is the process of visually inspecting the surface of a piece of equipment. This method is the simplest method of NDT.

Advantages:

- — Low cost;
- — Does not require a high level expertise or training;
- — No equipment required.

Limitations:

- — Surfaces available to inspect, be clean and uncoated;
- — Only surface indications can be detected;
- — Only macroscopic indications can be detected.

Several standards have been developed that provide guidance on performing visual inspections. Further guidance on visual inspections can be found Refs [42-44] see Appendix VIII).

6.1.5.4. *Ultrasonic thickness testing*

Ultrasonic thickness (UT) testing uses a transducer to send high frequency waves into a test piece. The time required for the sound waves to travel through the test piece and be reflected by adjacent surfaces is measured and used to determine the thickness in that location. A change in measurement indicates a change in thickness or that a defect is present. This method can be used on a range of materials if the speed of sound within that material is known.

Ultrasonic thickness testing is particularly useful to detect internal corrosion of the walls of pipes, tanks, pressure vessels, and other equipment. General thickness testing can be used on parts of uranium processing plants such as grinding mill shells, crusher liners and structural beams where loss of thickness may have occurred.

Advantages:

- The location of the defect within the test piece can be determined;
- Access is required on one side only to take measurements;
- Preparation is minimal;
- Can be used on a range of materials;
- UT equipment is inexpensive.

Limitations:

- A trained and certified technician is required to complete the testing and analyse the data.
- Surface coatings and finish on equipment can affect the accuracy of readings.
- The entire area is required to be scanned, meaning that the analysis and interpretation of results can be time consuming.

Standards have been developed that provide guidance on performing UT testing. Further guidance on UT testing can be found in Refs [45-48] (see Appendix VIII).

6.1.5.5. *Rapid motion scanner*

A rapid motion scanner (RMS) test is based on UT technology but uses a remote controlled crawler to detect the thickness over a larger area. The data is analysed via software that maps the thickness data to create a high resolution map of the area scanned. The colour ranges on the software can then be adjusted to help detect areas which are below the minimum allowable thickness. The crawler equipment uses magnetic wheels to attach itself to the wall of a tank or pressure vessel and gradually moves over the surfaces while the probe is moved back and forth. Different sized scanners are available for different applications.

Rapid motion scanners can be used to scan the external shell of a lined tank to show any internal corrosion within the tank or behind the internal liner. This method can be used in lieu of removing the internal liner to complete an internal inspection on the shell.

Advantages:

- Creates a colourized map of the surface;
- Retains meta data, which can be tuned to assist detecting required thicknesses;

— Can scan large areas, with the largest equipment having the ability to scan a 600 mm strip at a time.

Limitations:

— The equipment needs to be tethered to prevent it from falling from vertical surfaces.
— Relatively slow moving.
— High water consumption (as a continuous a flow of water is required as a bridge between the surface and the sensor).

The previously referenced standards for UT (see Section 6.1.5.4) also apply to RMS.

6.1.5.6. Eddy current testing

Eddy current testing utilizes electromagnetic induction to detect surface and near surface indications. This technique is useful for completing inspections on painted surfaces. Painted welds can therefore be inspected for external pressure vessel and tank inspections without the need to remove the paint.

Advantages:

— Sensitive to small cracks;
— Wide range of applications;
— Detects indications through paint;
— Inspection of complex shapes;
— Probe does not need to contact the surface.

Limitations:

— Only used for conductive materials;
— Requires trained and skilled technicians;
— Only surface and near surface flaws can be detected;
— Test reference standards are required to set up each inspection.

Several standards have been developed that provide guidance on performing eddy current inspections. Further guidance on eddy current inspections can be found in Refs [49-53] (see Appendix VIII).

6.1.5.7. Saturated low frequency eddy current

The saturated low frequency eddy current (SLOFEC^{TM} [5]) technique is a screening tool that uses eddy current technology, in combination with a magnetic field, to detect loss of metal thickness and perform corrosion mapping of the floor of a tank. This technique can detect corrosion on

---

[5] SLOFEC^{TM} is a registered as a trademark by Konroll Tecknik GmbH, and further reading on this technique can be found on the website http://www.kontrolltechnik.com/methods/slofec.

either side of the floor plate (i.e. topside and underside). For coated or rubber lined tanks, this method is ideal, because the scan can be carried out through the coating or rubber lining.

Floor thickness can be measured on a grid pattern of six readings per plate. The annular ring thickness can be measured at twenty equally spaced locations around the tank circumference.

Advantages:

— Allows scanning through the coated or rubber lined floor up to 10 mm thickness;
— Relatively rapid scanning tool;
— Can negotiate uneven surfaces;
— Can differentiate between topside and underside defects;
— Can be used to scan the floors of stainless steel tanks.

Limitations:

— It is only a screening tool and any identified defects may require verification using manual UT.
— The testing equipment is not able to access the shell to floor joint area (the critical zone for corrosion failures) in the vicinity of the butt weld. The restrictions are due to the specific design of the SLOFEC™ scanner, position of the sensors and the geometry of the device. The inaccessible area is commonly referred to as the 'dead zone'. Because the 'dead zone' is located within the tank's critical zone, alternative methods to scan the critical zone need to be employed to verify its integrity (e.g. in situ UT).

The previously referenced standards for eddy current testing (see Section 6.1.5.6) also apply to saturated low frequency eddy current testing.

6.1.5.8. Dye penetrant inspection

Dye penetrant inspection (DPI), also known as liquid penetrant inspection, is used to detect any surface discontinuities. This method can detect cracks, porosity and flaws on metal surfaces and is particularly well suited to weld testing. This inspection method uses a dye (often red) which is sprayed or brushed on to a surface where the dye penetrates the discontinuity through capillary action. The excess dye is wiped off the surface and a developer (usually white) applied over the same surface. The developer attracts the dye to the surface and enables the condition monitoring technician to observe any flaws. Fluorescent dye can be used in conjunction with a fluorescent light to more easily detect the discontinuity, which is particularly useful in low light applications, for example within a pressure vessel.

Advantages:

— High sensitivity;
— Suitable for large range of materials and complex shapes;
— Discontinuities are created on the surface for easy defect identification and further examination;
— Low cost;
— Very portable (spray can application).

Limitations:

- — Only detects surface discontinuities;
- — Defects under the surface cannot be detected;
- — Only non-porous surfaces can be tested;
- — Pre- and post-cleaning of surfaces is required.

Standards have been developed that provide guidance on performing DPI. Further guidance on DPI can be found in Refs [54, 55] (see Appendix VIII).

6.1.5.9. *Acoustic emission testing*

Acoustic emission testing is used to detect cracking within pressure vessels. Highly sensitive sensors are placed over the surface of a vessel. Once the vessel is pressurized, small cracks within the vessel can be detected and the location of the crack is triangulated from sensors to pinpoint the exact location. A map of the vessel can then be created to show the location and severity of defects. This method is particularly suited to ammonia storage vessels which are prone to stress corrosion cracking.

Advantages:

- — In service inspection allows equipment to remain online;
- — High sensitivity;
- — Cost effective compared to internal NDT.

Limitations:

- — Limited availability of equipment and trained personnel to perform testing;
- — More sensors required for larger equipment;
- — Difficult to use in noisy work environments;
- — Results are indicative only and require other NDT methods to perform more thorough examination.

Several standards have been developed that provide guidance on performing acoustic emission inspections. Further guidance on acoustic emission inspections can be found in Refs [56-58] (Appendix VIII).

6.1.5.10. *Magnetic flux leakage*

Magnetic flux leakage (MFL) can be employed as an alternate method to SLOFEC™ to scan the floor of tanks manufactured from carbon steel. The technique uses permanent magnets and sensor coils incorporated in the mobile scanner that can be moved across the floor of the tank. The permanent magnets are used to induce a magnetic field, while the sensor coils detect 'leaks' (or disturbances) in the resulting field which originate from defects in the metal. The amplitude of the leaking field is measured by the sensor coils and compared to known defects to predict (or infer) wall losses.

MFL is able to perform a scan through coating and rubber lining, although there are restrictions on the maximum thickness of the coating which the scanner can measure through, which is typically lower than that of SLOFEC™. Therefore, MFL is not normally suitable for the scanning of thicker rubber lined floors.

Advantages:

- It is a rapid scanning tool;
- It can discriminate between topside and underside defects;
- Various signal parameters allow clear distinction between weldments and actual defects.

Limitations:

- It can only scan the floors which can be magnetized (i.e. carbon steel tanks);
- False indications could result from poorly cleaned and prepared surfaces containing sand, scale and heavy rust;
- It provides inferred plate thickness measurements, which need to be verified by manual UT;
- Similar to SLOFEC™, the inaccessible 'dead zone' around the circumference of the tank adjacent to the floor-to-shell weld needs to be scanned using alternative means (e.g. manual UT).

Detailed guidance on performing magnetic flux inspections is available in Ref. [59] (see Appendix VIII).

6.1.5.11. Radiography

Radiography uses portable X ray equipment or gamma ray sources to create an image of the equipment on a film. This method is used for weld examination however is becoming less popular due to safety restrictions, cost and alternate more modern technologies such as PA.

Advantages:

- Image produced can show good detail of the welds.

Limitations:

- Quality of image dependent on set up of equipment;
- Only black and white images are produced;
- Images are required to be developed;
- Images are not digital and need to be rescanned;
- Radiation protection arrangements require an exclusion zone around the radiography being performed;
- High cost.

Several standards have been developed that provide guidance on performing radiography inspections. Further guidance on radiography can be found in Refs [60-62] (see Appendix VIII).

6.2. STEEL INFRASTRUCTURE

Steel structures in facilities intended for permanent or regular human and equipment occupancy ideally have a documented maintenance regime to ensure their continued integrity. Defects might be introduced through inadequate design, manufacturing or construction before the structure is placed in operation, and less often due to inappropriate operational functioning or maintenance induced defects. Deterioration of steel infrastructure or assets is normal, mainly due to ageing and weathering. However, due to the corrosive nature of chemicals used in uranium ore processing (e.g. sulphuric acid), accelerated levels of structural deterioration may be observed in the steel structures of uranium processing facilities, when compared with other minerals extraction plants. During the life cycle of a uranium production facility, structural assessments need to be performed. Furthermore, it is necessary to perform a structural assessment after a major natural event (e.g. tornado, hurricane, earthquake).

The purpose of structural assessments is to ensure structural asset integrity for both safety purposes and operational efficiency. The assessment will document the structural condition, establish the structural serviceability, enable planning for maintenance or repair and communicate compliance matters as per national and local regulations. A comprehensive structural assessment typically includes a review of documentation, component classification, field investigation, testing, analysis and a detailed report. Structural inspections may include one or more of the following activities:

- Visual observation;
- Thickness measuring;
- Field and laboratory testing;
- Non-destructive examination;
- Numerical or software based analysis.

Typically, the maintenance department of a uranium mine and processing facility is responsible for completing a functional inspection of all the structures, for non-substantial repairs or maintenance activities. The structural inspection will normally be escalated to a professional civil or structural engineer if the inspection identifies issues that require a specialist assessment to interpret the visual finding or other structural concerns (e.g. significant cracks, corrosion or erosion).

The structural infrastructure of a facility can be inspected by area and prioritized in terms of structural integrity concerns. Good practice would be to approach the structural inspections using a risk based inspection methodology. The frequency of inspections is often linked to the risk classification. For example, high risk category structures might be thoroughly inspected every two years.

All structural elements require structural integrity inspections such as pipe supports, lighting elements, walkways, handrails, etc. Inspection requirements may also be legislated in certain countries or regions. Tank supports and attachments are often considered structural elements and inspected separately from the tanks. It is good practice to continuously monitor the corrosion of lower risk defects for gradual degradation.

6.3. CRANES AND LIFTING EQUIPMENT

The ACA for fixed and mobile cranes used in the mine or processing facility will be determined by the type and size of crane and as required by nation and local regulations. Cranes are typically inspected annually by competent personnel. In addition, all lifting equipment (i.e. chains, slings, hooks, shackles, etc.) are typically inspected every three months or as required by national and/or local regulations. A useful reference guide on crane inspections can be found in Refs [63, 64] (see Appendix VIII).

6.4. VENTILATION SYSTEMS

Ventilation systems in underground uranium mines and uranium processing facilities are critical assets that require routine inspection and, as necessary, subsequent maintenance. Ventilation systems control the levels of airborne contaminants in working areas, as well as in process vessels and production equipment. The management of radon gas, hydrogen off gas during acid leaching, sulphur dioxide from acid production plants, calciner or roaster off gassing and uranium dust particulate matter are all concerns specific to the uranium industry that require special consideration.

A well designed system will have an air exchange rate that will maintain the air quality within the mine and/or processing facility to ensure that it is safe for workers in accordance with national or local industrial hygiene regulations. The system design needs to be considered during the design phase of the facility or mine, and the ventilation specifications need to be analysed by a competent engineer. The documented design specifications needs to include the required ventilation duct sizes, materials of construction, air flow rate, air filter or scrubber system requirements, and locations of inlets and outlets. A useful guide for ventilation design is provided in Ref. [65].

Maintenance of an installed ventilation system in a uranium mine and processing facility is imperative and cannot be overlooked. A routine maintenance plan for a ventilation system would typically include, at a minimum, the following, occurring every six months:

- Visual inspection of the ducting exterior, checking for obvious signs of damage, holes, or corrosion;
- Velocity measurements of the air flow within the ducts at specified locations, to determine if the system is functioning as per original design;
- Air filter or scrubber system cleaning;
- Electric motor inspection for fan drives, which may include amperage measurements;
- Fan blade cleaning and visual inspection for wear and damage.

It is industry good practice to keep the results of all the inspection data over an operation's lifetime, specifically the velocity measurements are expected to be loaded within the CMMS. Ventilation inspection and performance data can be presented graphically and any trends over time can be evaluated. If there is a significant change from one measurement to the next (e.g. a 10% reduction), then it may indicate that a component within the ventilation system needs further inspection or investigation. For instance, if there is a low air flow velocity in a duct, it may indicate that there is a blockage that needs to be corrected or that there is a fan that is not functioning to its design capacity.

The materials that the ventilation system is constructed from need to account for the location where the ducting will be installed. There are many situations in a uranium processing facility where the local environment may corrode ducting at an advanced rate. In addition, static electrical charge can build in a duct system, so it is industry practice to electrically ground the ducting. The original design specifications need to provide a detailed description of the correct materials for maintenance applications. If a portion of the ducting is to be replaced, it is good practice to replace the section with the material specified, or in some cases, a more advanced material. Care is needed not to replace with the lowest cost alternative, as this may lead to unintended hazardous consequences due to system underperformance and premature or unexpected failure.

During maintenance activities, temporary ventilation may be required to enter some areas of the plant, especially for confined spaces, such as tanks, or process equipment. During this type of maintenance work, measures are required to ensure the protection of workers. Many States have regulations for enclosed spaces which are required to be complied with. Typically, this would include constant monitoring of the atmosphere inside the confined spaces, and an independent worker (i.e. a spotter who is not performing maintenance activities) outside the vessel to verify safe work conditions and initiate emergency response procedures.

6.5. STORAGE TANKS

This section provides guidance of the extent, frequency and methods of both prescriptive and risk based inspection and monitoring activities, to maintain the integrity of above ground storage tanks. In this section, industrial storage tanks, for example those used to manage bulk commodities (e.g. fuels, process chemicals), and processing tanks, where the chemical processes in uranium production take place (e.g. blending, leaching, extraction) are both considered. Maintenance activities for the auxiliary equipment of storage tanks (e.g. agitator, motor) are excluded from this section, however this is included when developing maintenance tactics through the ACA and FMEA processes.

Examples of commonly used standards for the construction and maintenance of storage tanks are Refs [42, 66-69] (see Appendix VIII).

Storage vessels within a uranium processing facility are typically not registered assets (i.e. unlike pressure vessels). However, many tanks within a uranium processing facility contain hazardous chemicals or substances. Failure of tanks can disrupt normal production and, in some cases, may pose a risk of injury to workers, the public and environment. The requirements for storage tank inspection and monitoring activities are therefore primarily risk based.

The following subsections provide simplified guidance on maintenance inspection and repair. More detailed tank inspection guidance is provided in Appendix VI.

6.5.1. Asset criticality assessment

Tank details and specifications need to be understood and identified, including dimensional data, materials of construction, lining materials and fluid contents, year of installation, results of inspections and resulting maintenance repairs, for all storage tanks within the mine and processing facility. In addition, an ACA is to be completed for all tanks and this needs to consider the service condition, design, operating history and dangerous goods classification of the stored material. The risk associated with the tank will determine the inspection frequency, with higher risk tanks requiring more frequent internal and external inspections. To determine the risk level, the consequence of failure for the credible damage mechanisms and likelihood of failure ought to be taken into consideration.

If prior inspection data is not available, it is difficult to estimate the likelihood of failure. In this case, a consequence based assessment is expected to be carried out to prioritize inspections with the highest consequence assets being inspected first. Once the inspection data has been gathered, the likelihood of failure, based on the condition, can be used in conjunction with the consequence to calculate the risk. Tanks can then be grouped together based on their consequence or risk level and have maintenance plans assigned.

6.5.2. Damage mechanisms and failure modes

It is good practice to define damage mechanisms for each tank (similar to the FMEA process) so that targeted inspection programmes can be developed (see also Appendix V). In the absence of having the damage mechanisms defined, a general approach for inspection is taken, which requires an increased reliance on a qualified tank inspector and the operation's engineering team.

Industrial storage and processing tanks are designed to tolerate significant head pressure and are designed on a 'leak before failure criteria'. However, due to the corrosive nature of uranium hydrometallurgical processes, a small leak may lead to major tank failure in a relatively short time frame. A select number of tanks within the uranium processing facility may be considered critical risk because of their size, contents and risk to personnel and the surrounding environment if the tank fails.

The highest risk location for mild steel tanks is on the floor plate within 75 mm of the floor-to-shell weld. This is referred to as the 'critical zone'. Figure 10 illustrates the location of the critical zone in storage and processing tanks. This location, being close to the perimeter of the floor plate, is particularly susceptible to under floor corrosion due to ingress of moisture under the floor of the tank. Under floor corrosion at this location, coupled with high head pressure from within the tank can result in catastrophic failure of the tank if the weld seam fails. For this reason, it is important that the critical zone is inspected using NDT techniques and the seal around the base of the tank between the floor plate and plinth needs to be maintained. For rubber lined tanks, it is possible to measure the floor plate thickness through a rubber lining using SLOFEC™ (as described in Section 6.1.5.7). However, the SLOFEC™ equipment has limitations due to the physical location of the sensor on the equipment and is unable to take readings adjacent to the shell plate within the critical zone (the 'dead zone'). In consultation with a qualified tank inspector, it may be possible to remove small sections of the rubber lining adjacent to the shell, at selected sample points, to take manual UT readings of the floor, without having to remove the entire rubber lining of the tank floor.

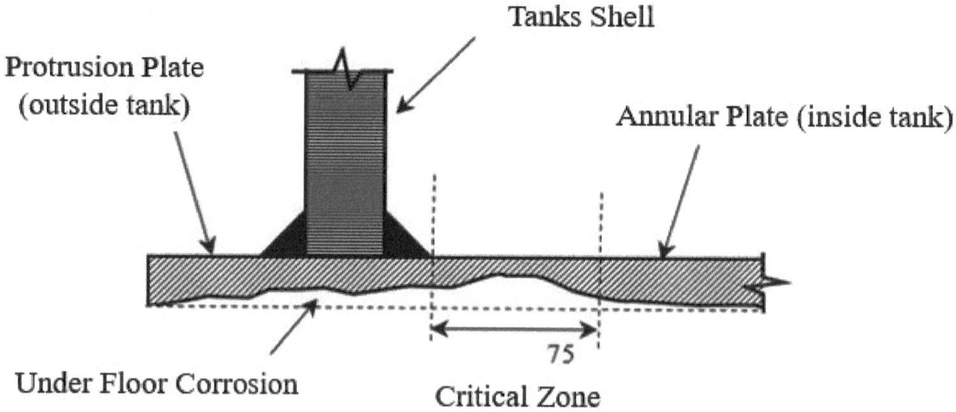

FIG. 10. Underfloor corrosion and critical zone.

Tank bunding (or berming) needs to be designed in accordance with national and local regulations and standards. Typically, this is 110% volumetric capacity of the largest tank within a bunded area, but may also require specific construction materials, techniques and physical design parameters. Draining tanks into bunds is not good practice, however it is sometimes unavoidable, particularly if there is a defect and the tank needs to be emptied quickly. For this reason, bunds need to be regularly inspected and all joints sealed to ensure it can hold liquid. A polymer spray-on liner can be applied to the bund to make it watertight if required.

Common failure modes of rubber lined mild steel storage tanks are summarized in Table 11. Additional detail on the inspection methods within Table 11 are outlined in Section 6.1.5.

TABLE 11. COMMON TANK DAMAGE MECHANISMS

Failure mode	Causes	Inspection method
Internal bottom corrosion	Breakdown in internal coating, erosive slurry, localized corrosion under deposits and at welds/crevices, stagnant or low flow areas, localized erosion at foreign objects and tank appurtenances (e.g. sparge supports, high caustic concentration and temperature, traces of water in diesel fuel accumulated at the bottom of the storage tank)	VI, SLOFEC™ (rubber in place)
Under bottom corrosion	Under bottom moisture from liquid overflows, tank bottom leaks, poor drainage around tank, high groundwater, tank settlement, poor foundation materials (e.g. saline, coarse)	VI, UT, SLOFEC™, MFL
Internal coating damage	Erosive slurry and/or carbon, poor application (e.g. quality control or environmental conditions), physical damage, osmotic blistering, ageing, damage from equipment during inspection	VI, Thermography (external inspection during operation)
Internal shell corrosion	Breakdown of internal coating, erosive slurry and/or carbon, localized erosion at tank appurtenances (e.g. baffles)	VI, UT, RMS, thermography (external inspection during operation)
External corrosion	Breakdown in external coating, external moisture (e.g. precipitation or overflows, external salt or pollution, overflow)	VI, UT, RMS, thermography (external inspection during operation)
Corrosion under insulation	Damaged insulation, moisture ingress, breakdown of external coating	VI, UT
External coating damage	Ultraviolet degradation, poor application, ageing, mechanical damage to coating	VI
Physical damage	Equipment failure, human error, natural disasters	VI
Cracking	Thermal fatigue from exceeding operating temperature, brittle fracture, thick sections (>12 mm) and low temperatures	MPI
Foundation	Soil settlement, poor foundation drainage	VI, settlement survey

6.5.3. Frequency of inspection (external and internal)

The inspection frequency for a tank is expected to be aligned with the criticality assessment for the tank. An internal inspection frequency needs to be assigned to the tank based on prevalence of damage mechanisms and their calculated remaining life. These processes and calculations are outlined in detail in Ref. [42], which is considered an important industry standard for maintaining and repairing tanks.

An external inspection of the tank and its components is most commonly double the time interval of the internal inspection frequency. External inspections include NDT over the external shell, roof and nozzles and also a visual inspection by a certified tank inspector.

An example of an inspection frequency table used within a uranium processing facility is shown in Table 12.

TABLE 12. EXAMPLE OF TANK INSPECTION FREQUENCY FOR EACH RISK GROUP

Inspection method	**Risk group**			
	Critical	**High**	**Medium**	**Low**
Visual inspection	Monthly	6 monthly	NA	NA
External visual and NDT	Annual	2 yearly	4 yearly	5 yearly
Internal visual and NDT	Annual	4 yearly	8 yearly	10 yearly

The internal inspection interval does not normally exceed ten years following installation. The initial internal inspection period can be calculated from the tables provided in section 6.4.2.1 of Ref. [42]. Subsequent internal inspection intervals can be calculated from section 6.4.2.2 of Ref. [42]. Quantitative measurements of the tank floor thickness are required to determine the tank floor integrity when performing internal inspection.

Reference [42] recommends that external inspections of tanks are to be conducted at least every five years. However, after sufficient quantitative data is obtained to predict the rate of tank wall deterioration, the inspection frequency can be reviewed. Section 6.3 of Ref. [42] describes the methods to calculate an inspection frequency.

An external visual inspection interval and method needs to be established for higher risk tanks. Typically, the inspection can be completed by trained and qualified production operators with the condition reported to the engineering team for assessment and further action if required.

6.5.4. Obsolete tanks

Obsolete tanks ought to be managed as a structure and be included in the structural integrity assessments for the plant (see Section 6.2). Obsolete tanks need to be drained and cleaned, kept in a locked open position and fully isolated from other possible process inflows.

6.5.5. Critical rubber lined tanks

Within uranium processing facilities, rubber lined tanks are commonly found in areas for leaching, neutralization, clarifier and backwash processes. Large rubber lined processing tanks (e.g. greater than 5 metres in height) and tanks in elevated positions are generally classified as critical assets, as they have a high risk of failure due to the integrity of the tank being reliant on the integrity of the rubber liner. The rubber liner can be damaged or fail for a number of reasons which can then rapidly lead to corrosion of the mild steel shell causing catastrophic failure of the tank. Appendix VII identifies the damage mechanisms and controls to mitigate the risk of rubber failure to as low as reasonably practical. These damage mechanisms, impacts and controls are unique for each uranium processing facility and need to be assessed accordingly.

6.5.6. Fibreglass composite tanks

Fibreglass composite tanks and vessels may be selected for use where steel would otherwise corrode due to contact with corrosive chemicals used in uranium processing (including process waste streams). There are a large variety of resins and fibreglass combinations that may be selected to best suit the design parameters, with consideration for contact with the process solutions.

Fibreglass composite tanks are typically drained, cleaned, and inspected on a yearly basis. The internal layer of resin that is in contact with the process solution may erode over time, thereby requiring close visual inspection at a prescribed frequency. The internal layer of fibreglass that is in closest contact to the process solution, is ideally designed to be a different colour than the rest of the fibreglass layers. This allows the inspector to observe when the erosion of the vessel reaches this layer, indicating that the fibreglass composite vessel requires repair. Most fibreglass vessels can simply be cleaned, and then repaired by rebuilding the wear surfaces to original thickness using a combination of resin and fibreglass matting. An additional consideration for inspection includes identifying any cracking of the vessel, which tends to occur at connection points with piping, or at locations where a mixing agitator or similar mechanical device is mounted.

6.6. PRESSURE VESSELS

Pressure vessels are categorized as high risk equipment as catastrophic failure of a pressure vessel can result in severe consequences to people and the environment. National and local regulations often require registration of pressure vessels to ensure they have an appropriate hazard level rating, installed pressure relief devices, a programme of maintenance inspection and repair, and other controls measures in place to manage and mitigate the risk (see Section 4.7). This registration may also allow for local authorities to conduct ad hoc inspections and reviews on pressure vessels, in addition to maintenance inspections performed by the owner/operator.

A systematic categorization and assessment of pressure vessels requires information about dimensions, materials, fluid description, operating pressures and any other criteria specific to the operation of pressure vessel equipment. This information is typically kept in a centralized database by the owner/operator of a uranium mine and processing facility (e.g. CMMS).

The assessment and management of pressure vessels may be subject to specific regulatory requirements and compliance with national standards, which generally follow a similar process. The vessel volume, contents and system pressure are used to specify a hazard level, which is designated by a lookup table or matrix from the relevant standard. The hazard level is then used to determine the requirements for internal and external inspections, including scope and the frequency of such inspections. For example, Ref. [70] can be used to identify the hazard level of a pressure vessel, and table 4.1 of Ref. [71] can be used to specify the correct maintenance intervals for that hazard level.

Further guidance on assessment and management of pressure vessels can be found in Refs [72-76] (see Appendix VIII).

6.7. PIPING

All piping within a uranium mine and processing facility is to be assessed using the ACA process outlined in Section 4.2.

Piping used to carry process slurries within bunded (or bermed) processing facility are usually classified as lower risk assets. This is due to the lower consequences associated with the failure of process piping as any leaks usually report to a bund (or berm). These leaks are quickly detected through production operator inspections and can be repaired. There are exceptions to this, where process piping may pass outside bunded (or bermed) areas between sections of the processing facility or are hidden within infrastructure or underground services.

High wear locations within piping systems may need to be inspected to mitigate against failure and reduce risks to production, health, safety and environment. This may include inspecting nozzles, elbows, reducers or other high wear locations. Due to the abrasive wear on pipes within uranium processing facilities, the lower portion of a pipe can experience increased wear. If this occurs, pipe rotations may be required at regular intervals (i.e. every six months to a year) to prolong the operational lifespan of the pipe, or replaced. This is often the case for tailings pipelines made from high density polyethylene (HDPE) which are more prone to erosion.

Maintenance activities may include pigging of pipelines, where a mechanical 'pig' is placed inside a pipe and movement of the pig through the pipe is accomplished by pumping process fluids or water through the pipe. The pig can be used for a range of maintenance tasks and for removing scale or blockages in preparation for inspecting pipes. Pigging requires the piping to be set up to install and retrieve the pig.

6.7.1. Common process liquor and slurry piping

There are three common types of piping used for uranium processing which are suitable for low pH process slurries and liquors. Each type has strengths and weaknesses. These are outlined in Table 13 and described in more detail below.

TABLE 13. COMMON PIPING USED IN URANIUM PROCESSING FACILITIES

Pipe	Resistance to wear	Failure after installation	Unsupported span length	Ease of modification	Shelf life	Cost
Rubber lined mild steel	High	High	Long	Medium	Medium	Medium
HDPE	Medium	Low	Short	Easy	Long	Low
Stainless steel	High	Low	Long	Difficult	Long	High

6.7.1.1. Rubber lined mild steel piping

Mild steel with rubber lining is the most common type of pipe used within a processing facility for slurries. Mild steel piping can be easily fabricated at an operations workshop facility; however, it requires a rubber liner to be installed to protect against acidic liquids coming in to contact with the mild steel pipe. The durability of rubber lined piping is high as the rubber lining is resilient to wear from abrasive slurries; however, the rubber lining can wear over time, particularly at elbows and ridges where there is increased turbulence in the flow. The reliability of a rubber lined pipe is often related to the quality of the installation of the rubber lining inside the pipe. If errors are made when the internal walls of the pipe are lined with rubber then the rubber lining can have areas of poor adhesion, can bubble or become brittle over a long period of time. These defects in the rubber lining can therefore result in a hole in the pipe caused by corrosion. Spare sections of rubber lined pipe enables the quick replacement of failed sections of piping. Measures are required to preserve the rubber lining inside of the pipe while the spare sections are stored, as the rubber can be damaged by environmental conditions (e.g. sun, rain), or through mechanical damage due to handling. Spare rubber lined pipes need to be wrapped, have blanks fitted to flange faces and stored undercover. If materials and expertise are available, mild steel piping can be rubber lined just prior to installation to prevent use of aged rubber and to ensure the quality of the rubber.

If any of the sections of rubber lined piping are exposed to organic chemicals such as those used in solvent extraction circuits, then organic resistant rubber lined piping is to be used. This is due to the ability of some organic compounds to readily break down (i.e. dissolve) natural rubber. Consideration ought to be taken when raffinate streams from the SX circuit, which may contain traces of SX organic compounds, are recycled back to leaching or the CCD circuit as this may impact the integrity of the rubber lining in the pipes over time. Stainless steel, fibreglass reinforced plastic or HDPE piping is more commonly used in solvent extraction processes.

More information on rubber linings is provided in Section 6.9.

6.7.1.2. Stainless steel piping

Stainless steel piping is the most chemical resilient and reliable option for hydrometallurgical processing facilities; however, it has a significantly higher cost compared to rubber lined mild steel or HDPE pipes. Stainless steel is also more difficult to modify or repair. Another limitation of stainless steel is its propensity for fatigue fracturing in areas of high vibration and cyclical stresses. Stainless steel can also wear at high turbulence areas creating a gouge in the pipe. This defect can occur at small step heights (i.e. 5 mm), typically at pipe joints, where valves or other attachments exist in the pipeline or when transitioning from stainless steel to rubber lined mild steel. Spare sections of stainless steel piping can be stored indefinitely and easily installed by maintenance tradespersons, without the need for specialist rubber lining or poly welding equipment, both of which are often difficult to source. Although stainless steel piping is more expensive, some uranium processing facilities use it almost exclusively due to its improved reliability.

6.7.1.3. High density polyethylene

High density polyethylene piping is the lowest cost piping material for process slurries and liquors. HDPE can be welded in long sections with a poly welder, but this type of piping is more commonly flanged as it is difficult to complete poly welding of pipelines in elevated positions (e.g. not at ground level). In addition, having flanged sections also allows for pipe sections to be easily exchanged. For small diameter piping up to 250 mm diameter, electrofusion couplings can be used to join HDPE piping, however the integrity of these joins are not as strong as poly welding, being more likely to fail. HDPE sections are prone to internal erosion, particularly at elbows or areas of high turbulence. Due to the flexibility of HDPE piping, it is not possible to span unsupported lengths of HDPE piping and it requires additional piping supports. For this reason, it is often run in cable trays to add additional support to the pipe.

HDPE piping can be inspected for internal erosion using UT scanning. The ability to test the pipeline wall thickness allows wear rates to be calculated and tracked. When performing NDT inspections, temperature variations on the external surfaces and internal surfaces of HDPE pipes need to be considered. It is necessary to calibrate the UT equipment for the temperature of the pipe, otherwise large errors can occur in the test results.

Spare HDPE piping can have a long shelf life when stored correctly and away from direct sunlight.

6.7.2. Inspection frequency

The assessment and categorization of piping for inspection is usually guided by national and local regulations, however the process generally follows a similar assessment process to that of pressure vessels (see Section 6.6). The hazard level and assessment of all piping ought to be completed to determine a suitable inspection frequency, using national standards or good practice industry guidelines. For example, Ref. [70] can be used to determine the hazard level and inspection frequency of piping, when used in conjunction with Ref. [71].

To simplify the process, piping with similar service and sizing ranges can be 'grouped' where these have similar hazard levels. Once these groups are established, a risk assessment can be completed on piping groups with consideration of operational based risk factors, to determine the hazard level associated with the pipe and subsequently inform the inspection regime. This assessment process is expected to also identify any piping that may require additional inspection (e.g. piping which goes outside a bunded or bermed areas or over a walkway).

Good industry practice is to develop a piping register that classifies the piping by location, contents and hazard level, and defines inspection intervals in accordance with the relevant standards. An example of a uranium processing facility piping register, for a set of piping assessed to Ref. [70] and then adjusted through a risk based inspection (RBI) for localized conditions, is outlined in Table 14.

TABLE 14. EXAMPLE OF A PIPING INSPECTION INTERVAL (ADAPTED ROM REF. [70])

Service	**Piping diameter (mm)**	**AS4343 hazard level**	**Inspection interval (years)**			
			AS4343 external	**AS4343 internal**	**RBI external**	**RBI internal**
Anhydrous ammonia	80–100	B	2	4	2	4
	40–80	C	2	4	2	4
	25	D	8	20	8	20
Sulphuric acid	25–100	E	NA	NA	5	10
Petrol, kerosene	50–100	E	NA	NA	4	8
	100–250	D	8	20	2	4
Distillate, fuel oil	150	D	8	20	2	4
	25–100	E	NA	NA	4	8
Tailings	90–150	D	8	20	4	8
Process water tailings	300–450	B	2	4	2	4
	200–300	C	2	4	2	4
	150	D	8	20	2	4

Table 14 shows that according to Ref. [70], there is no inspection required for sulphuric acid, petrol (piping 50–100 mm) and distillate (piping 25–100 mm). However, the risk based assessment method identified that some inspections need be performed due to a localized condition. This could be for several reasons including proximity to people, the potential effect on the environment or licence to operate requirements.

For further information, hazard levels are described in section 2.1.1 of Ref. [70] (and captured in Table 14), are summarized in Table 15.

TABLE 15. PRESSURE EQUIPMENT HAZARD LEVELS (ADAPTED FROM REF. [70])

AS 4343 hazard level	Definition	Example
A	High	Large vessels (i.e. 4000 t ethane / 7000 t butane / 25000 t ammonia / 200 t chlorine vessels)
B	Medium	Most shop fabricated boilers and pressure vessels
C and D	Low and extra low	Small pressure equipment or equipment with low hazard contents
E	Negligible	Usually exempt from special regulatory control but covered by general plant safety regulation

6.7.3. Inspection methods

To determine piping inspection methods, typical damage mechanisms and failure modes need to first be determined (see Section 4.4). The predominant modes of failure in pipeline systems occur due to general or localized loss of wall thickness caused by internal and external corrosion, as well as erosive wear.

Designated condition monitoring locations need to be established on piping with consideration for the pipe operating conditions and the nature of fluids being transported. Monitoring would normally be in the locations most susceptible to the damage mechanisms identified.

External inspections of piping include visual examination of the condition of the piping, fittings, flanges, flange gasket faces as well as any free standing pipe support components such as hangers, guides and shoes, for general erosion or corrosion. NDT examination, using techniques such as UT testing, can be used to measure the remaining thickness of structural piping supports, to measure the pipe wall thickness and to determine the extent of corrosion or damage.

Ultrasonic thickness testing for wall thickness is considered the equivalent of an internal inspection as is detects wall loss on the inside of the pipe, particularly for small diameter pipes. The inspection points are performed at the top, bottom, both sides of the pipe at lengths along the pipe, and at pipe elbows at the entrance, exit and partway around the bend.

Further guidance on piping inspection methods can be found in Refs [77-79] (see Appendix VIII).

6.8. PRESSURE SAFETY VALVES AND PRESSURE RELIEF VALVES

Pressure safety valves and PRVs are used on systems which are, or can become, over pressurized. Examples of pressurized assets include boilers, pressure vessels, compressed air, ammonia distribution systems, gas and some piping systems. Piping and vessels can catastrophically fail due to being over pressurized, with potentially catastrophic consequences to people, the environment and other equipment in proximity. Pressure safety valves are often the final safety control measure before catastrophic failure of a pressurized system; therefore, it is essential that PSVs are managed as critical equipment and well maintained.

There are three main types of pressure relief devices:

— Pressure safety valve: PSVs are typically used for emergency situations where immediate pressure relief is required. These valves often have directional discharges which allow them to reach full relief flow quickly and remain open until a safe pressure is achieved.

— Pressure relief valve: A PRV will usually open at a slower rate once the discharge pressure is reached.
— Burst disc: Burst discs are perforated discs which burst open at a predetermined pressure. Burst discs are inexpensive to purchase and install, are leaktight, have an immediate response time, high flow area and ease of maintenance. However, once the disc has burst, the equipment it is protecting needs to be taken offline until the burst disc is replaced. Due to the irregularity of discs bursting, spare burst discs may not be within their certification date and are often difficult to source at short notice. Burst discs also rapidly discharge the pipe or vessel contents to the surrounding area, which can present a safety risk to personnel in proximity, depending on the contents and pressure.

For purposes of managing traceability, maintenance and replacement, it is best to treat PSVs, PRVs and burst discs in the same way.

A register of all pressure relief devices in use, along with dimensional data, serial number, set pressure, servicing requirements and other relevant information ought to be kept, usually within the CMMS.

All pressure relief devices need to be replaced as per national and local regulations and standards. For pressure relief devices protecting a vessel, this replacement interval can usually be aligned with the pressure vessel internal inspection, when the vessel is required to be taken offline for inspection and replacement of critical components.

Table 16 provides an example of the typical overhaul intervals for pressure relief devices for various types of services and equipment.

TABLE 16. PRESSURE RELIEF DEVICE CHANGE OUT FREQUENCY (ADAPTED FROM REF. [71])

Service or equipment	Frequency of change out
Boilers, corrosive service	Annual
Steam systems, deaerators, oil and tar or other viscous services, water	Every 2 years
Chlorine and soft seated valves	Every 3 years
Air receivers, inert gasses, petroleum and ammonia	Every 5 years

Typically, all pressure relief devices require a commissioning inspection, with the exception of burst discs which are designed to catastrophically fail and additionally are supplied with a commissioning certificate. Visual inspections by a competent person (usually a trained and qualified maintenance practitioner or engineer) are required for PSVs and PRVs and are specific to each location. The scope and frequency of each visual inspection is dependent on the liquid within the vessel or pipe and the immediate surrounding environment of the pressure relief device (e.g. if it is located near materials handling or processing equipment it will be more likely to become dirty and clogged, thereby requiring more frequent inspections and cleaning). An example of this is having annual inspections for pressure relief devices in ‘clean areas’ and six monthly inspections for pressure relief devices in ‘dirty service’ areas. For further reading, an example of good industry practices in inspection criteria for pressure relief devices is provided in section 4.7.3 of Ref. [71].

Further guidance on determining intervals for pressure safety valve testing and overhaul can be found in Refs [80, 81] (see Appendix VIII).

6.9. LININGS

Rubber lining is used within a uranium processing facility to protect against low pH process slurries and liquors that come in to contact with mild steel tanks, pump boxes, pipes and pressure vessels. Rubber linings are also used to reduce erosion in high wear areas such as on agitator blades and grinding mill inlet chutes. The integrity of the mild steel will remain as long as the rubber lining protects it. For acidic slurries, once the integrity of the rubber lining fails and process liquor comes in to contact with the mild steel, the steel can corrode through the full thickness of the steel in a matter of weeks.

Traditionally, the lifespan of a tank or pipe is calculated as the change in thickness of the steel wall or floor of the tank over time to reach the minimum allowable thickness. This calculation, supported by measurement, provides an indication of the remaining life of the asset. Thus, maintaining the integrity of the rubber lining is extremely important in optimizing the life of an asset in a uranium processing facility.

In some cases, scale can build up in areas of damaged rubber, having a secondary effect of protecting the lining. However, this is not to be relied upon to protect the steel or to prevent corrosion. As scale is typically more porous than the rubber lining, this generally only acts to reduce the rate of corrosion.

6.9.1. Types of rubber and their application

6.9.1.1. Natural rubber

Natural rubber is a common lining material used within a uranium processing facility for the grinding mills, leaching circuits, CCD and pyrolusite tanks. Natural rubber has a lower cost compared to organic resistant rubbers and is easier to apply due to its flexibility and adhesion. It is important to note that natural rubber is not resistant to kerosene and therefore it cannot be used within the solvent extraction process or any tanks where organic compounds may report to, such as the clarifier, neutralization tank or last CCD in the circuit (which may be able to receive organics from SX if required).

Due to the use of natural products (i.e. plant based) in the manufacturing of natural rubber, micropores can develop through the rubber. These pores can lead to water molecules being pulled through the rubber which can form in to 'rubber bubbles'. It is uncertain exactly how these bubbles form, however it is believed that the rubber glue, which is hydrophilic, pulls water molecules through the holes in the rubber. Due to the acid molecules being larger in diameter, they cannot enter into the bubble. The water which is contained within these bubbles is therefore usually almost pure. Due to a lack of oxygen behind the bubble, the rate of corrosion of the steel is very slow.

The size of bubbles and various other factors such as quality of rubber, glue application, temperature, process chemistry and other variables could potentially accelerate corrosion, therefore each operation will be different regarding management of bubbles. In general, the presence of small bubbles in rubber linings are not a reason for concern. However, small bubbles within proximity to each other can join up and form larger bubbles. Large bubbles or bubbles with pronounced edges can become damaged more easily and need to be monitored over time or removed.

The two main risks of bubbles to asset integrity are through a slow rate of corrosion through the tank shell, or the more significant risk that the bubble may tear or fail allowing acidic liquor to come in contact with the steel. Good practice is to cut out the sections of lining with bubbles and patch these areas during internal inspections and refurbishments, however this can also lead to maintenance induced failures if the rubber patches aren't applied properly. Section 6.9.2 describes techniques for rubber lining repair methods. A risk based approach is to be taken for bubbles and the decision to replace needs to be based on the size, shape, level of corrosion, speed at which the bubble has formed and the surrounding environment if a hole were to occur. Bubbles can be tracked using photographs to visually compare any changes since the previous inspection.

Another risk to asset integrity relates to the hardness of the natural rubber lining. If the rubber becomes significantly softer compared to original installation, the process solutions inside the vessel will erode the rubber lining at a higher rate, and significantly shorten the useful life of the rubber lining.

Natural rubber linings will soften and expand if in contact with organic petroleum solutions, requiring the lining to be replaced. An example of an undesired scenario is if solvent from SX flows back through neutralization or CCDs. However, if the organic solution can be removed immediately, and the tank rinsed with clean water, then the softening effect to the rubber will slow dramatically. The rubber may still be usable, however more frequent monitoring and inspection may be required.

The hardness of the rubber can be measured using a pin type durometer measuring gauge to take measurements over a large sample area and recording these measurements over a period of time (e.g. every six months for example). A trend can be evaluated to assess if the rubber can continue to provide protection to the tank shell or if maintenance repair to the lining is required.

6.9.1.2. Organic resistant rubber

Butyl rubber is resistant to polar and non-polar organics and can therefore be used as a lining within the SX plant or any areas of the processing facility which receives organics. Organic resistant rubber is stronger and does not have micropores like natural rubber, which means rubber bubbles do not form. However, organic resistant rubber is more expensive and difficult to apply, particularly for small diameters or patch repairs. In such conditions, stainless steel, fibreglass reinforced plastic or HDPE piping may provide a better alternative as these do not require rubber lining.

6.9.2. Rubber lining repair methods

6.9.2.1. Skive butt joint

A skive butt joint is the simplest of rubber joints and the quickest to apply (see Fig. 11). However, if the joint is not done properly, the overlap can become loose, particularly if the joint is facing into the flow of fluid (e.g. in an agitated tank). To add an additional layer of protection to the skive joint, a capping strip or layer can be added on top of the joint. This is shown on the right side of Fig. 11. For small sections that have been cut out of a rubber lining, the section may be filled using a skive joint and then a capping piece applied over the whole area to provide additional protection.

A skive with overlap is similar to a skive joint, however the inserted rubber lining overlaps the skive as shown in Fig. 12. This is the most common lining method as it is fast to install, provides more integrity then a skive butt joint and does not require a capping strip.

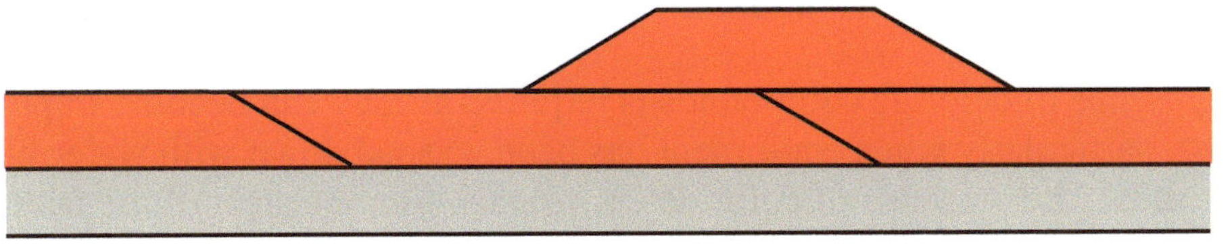

FIG. 11. Example of skive joint (left) and skive joint with capping strip (right).

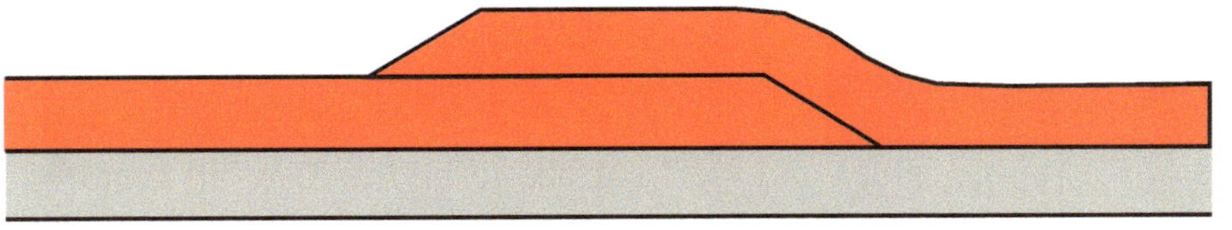

FIG. 12. Example of skive joint with overlap.

Repairs of rubber lining of tanks and vessels need to be completed in a way which minimizes the risk of joints becoming loose. This includes ensuring the leading edges of joints are not exposed to the flow of fluid.

A common area of failure is where stainless steel bolts, cleats or tabs are attached to the inside of a mild steel vessel or tank and are required to protrude through the thickness of the rubber (for example supporting upcomers in leach tanks or laterals within sand filters). The rubber seal surrounding the stainless steel tab can allow liquor to come in contact with the mild steel shell leading to corrosion. In these cases, it is good practice to weld stainless steel attachment plates to the inside shell such that the rubber lining is laid over the attachment plate. Any liquor which comes in to contact with the shell will therefore not cause corrosion.

The left side of Fig. 13 shows a bolt or cleat attached directly to the mild steel shell and the right side shows a stainless steel plate protecting the mild steel from ingress of liquid. In Fig. 13 the image on the right side also shows the attachment plate with chamfered edges. This is necessary for the rubber to adhere to the plate. Where a plate has been installed without chamfered edges, an epoxy or steel putty can be used to create a smooth profile for the rubber to sit against.

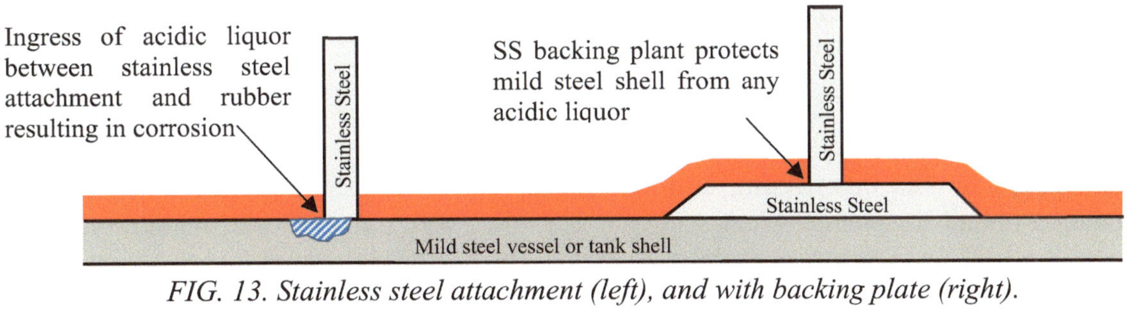

FIG. 13. Stainless steel attachment (left), and with backing plate (right).

6.9.3. Rubber installation quality control

During maintenance tasks requiring rubber installation, the manufacturer's instructions need to be followed closely to ensure the rubber lining serves the purpose of protecting the shell of the vessel or tank, pipes and other relevant assets.

6.9.3.1. Salt build up on steel prior to application of rubber lining

A check for the presence of chloride compounds is required as they can become trapped within mild steel surfaces, particularly when equipment has been washed down with water which may contain chlorides salts. The presence of chlorides in contact with the steel of the tank will lead to accelerated corrosion and therefore needs to be avoided. When installing rubber, it is important to check the surface is free from chlorides. Surfaces containing chlorides may need to be pressure washed or sand blasted prior to the rubber lining being installed.

6.9.3.2. Installation temperatures

The temperature of the rubber, glues and steel surface to be lined are all required to be kept within parameters specified by the manufacturer for effective rubber lining. Glues need to be stored in suitable environmental conditions (e.g. temperature and lighting conditions) and the expiry date of glues needs to be checked prior to use. The environment where the rubber is being installed may need to be air-conditioned, heated or dehumidified to ensure ambient conditions are within the parameters specified by the manufacturer. Heating tables can be used to warm rubber liner material to ensure the liner is at the prescribed temperature. Tanks may need to be covered with tarps or scaffold to allow the environment to be temperature controlled.

6.9.3.3. Surface preparation

The surface profile of the parent steel affects the ability of rubber to adhere to it. The surface profile, particularly after sand blasting or other abrasive preparations, needs to be checked to ensure the correct surface profile is obtained.

6.10. RESINS AND POLYMER LININGS

6.10.1. Glass reinforced plastics

In certain areas of a uranium processing plant, such as solvent extraction, rubber lining and hot work (i.e. welding, soldering, cutting, etc.) is not permitted due to the fire hazard of the solvent that is used in the SX circuit. As an alternative, glass reinforced plastics are often used as a lining material for plate work and pipes. This section provides an example of good practice for the specification, preparation and lining of steel and platework with glass reinforced plastics for corrosion protection for chemical duties compatible with vinyl ester resin (at temperatures normally up to 60 °C maximum).

6.10.2. Resin

The resins used for the main lining laminate are epoxy vinyl ester resins derived from the chemical compound ‘bisphenol A’. A seal coat is applied using a vinyl ester resin mixture and tinted with a clear dye to ensure complete coverage. The seal coat is cured fast to achieve a tack free hard cure within two hours of application. All methods of application are expected to conform to the resin manufacturer’s instructions.

6.10.3. Glass fibre reinforcement and finishing tissue

The reinforcement tissue used comprises of layers of 450 g/m^2 type E glass chopped strand mat having a resin binder. The finishing tissue used is generally a C-glass veil resin bonded tissue.

6.10.3.1. *Preparation*

The surface to be prepared needs to be abrasive blasted to standard grade Sa2½ (BS EN ISO 8501-1) finish all over and with a surface profile of 40 to 60 microns. Care is needed to avoid shadows when blasting.

The blasted metal needs to be cleaned to remove all dirt and dust.

6.10.3.2. *Priming*

As soon as possible after blasting, one coat of vinyl ester resin primer is applied to the steel surface at the rate of about 1 kg per 10 m^2. Alternatively, a fast curing vinyl ester resin primer can be applied at the rate of about 1 kg per 5 m^2.

6.10.3.3. *Main lining laminate*

The main lining laminate which results in total lining thickness of approximately +/- 6 mm is applied in the following sequence:

(1) Resin primer coat;
(2) One layer of C-glass tissue;
(3) Five layers of 450 g/m^2 chopped strand mat;
(4) One layer of C-glass tissue.

The laminating procedure for piping systems for uranium mining and processing facilities is expected to follow good industry practice and conform to relevant standards such as Refs [82, 83] (Appendix VIII).

The finish coat (in contact with the process solution) needs to be applied after pinhole testing the main laminate with a high voltage high frequency detector.

The finish coat consists of one layer of vinyl ester resin, paraffinated and tinted to ensure full coverage. It needs to be cured quickly to achieve an acetone-tack-free surface, of Barcol 31 hardness with 2 hours of application.

6.11. CONTROL SYSTEMS

Digital control systems (DCSs) in a uranium processing plant are used to monitor and control extraction systems and hydrometallurgical equipment. The DCSs also provides invaluable data on plant performance by which operations personnel can identify issues with asset health and maintenance requirements.

Digital control systems also require maintenance, such as cleaning of hardware, digital updates of software, maintenance of valve actuators and calibration checks of the instrumentation. This type of maintenance work is typically executed by specialist trained instrumentation technicians or technologists, with the assistance of trained information technology professionals.

In ageing facilities, the DCSs often require upgrade or replacement. Computer technology improves at a rapid pace and components that make up the control system, which may have been state of the art when installed, run the risk of no longer being available from suppliers in 10–15 years. This may include computers within the system that were originally installed with

16-bit or 32-bit processors, which no longer support software that has been designed for a modern 64-bit processors. Components that have been made obsolete by the OEM may still be sourced from third party vendors, until such time that market supplies run out.

Bandwidth availability on digital communication networks may also need to be upgraded to support the needs of modern DCSs. This may include increasing the wireless network capacity and capability to extend throughout the facility or installing fibre optic cabling throughout the facility to link modern computers and instrumentation components.

With consideration of the rapid advancement of digital communications, it is prudent to plan financially for replacement of digital infrastructure within the lifetime of an operating facility. New technology can be adopted in stepped increments, as complete changes of a DCS can be costly to implement.

Computer based systems play an essential role in uranium processing facilities, for safe and secure operations, storing and transporting nuclear material and other radioactive material, and maintaining physical protection systems. As such computer based systems need to be secured against criminal or intentional unauthorized acts [84]. Good practice is to use a digital firewall to isolate the DCS from open networks, with the intention to limit the ability for malware and hackers to access or take control of the processing control systems.

Redundant (or backup) DCSs are generally used to protect the facility in the event that the local control system stops working. This could take the form of having one centrally based computer (central control) with the ability to manage the entire plant operating system, while the local control system is being maintained or replaced. A digital backup of the control software needs to be saved. This is typically saved to a corporate server located off the site, so that in the event of a fire or other major unfavourable event the system can be rebuilt and restored to original state.

6.12. ELECTRICAL INFRASTRUCTURE AND BACKUP POWER SYSTEMS

Electrical infrastructure at a uranium mine and processing facility includes all electrical components downstream of the supplied high voltage incoming power, down to the lowest voltage electric plug. In addition, most facilities also have a source of backup power such as diesel generators capable of powering emergency lighting and safety systems or selected processing equipment, so that production can continue if there is a loss of power to the facility.

Electrical equipment requires a preventive maintenance plan which would include periodic cleaning, inspections, and testing to ensure it maintains integrity. As motor control centres and switch gears age, the electrical isolators might break down leading to arcing to ground, causing equipment damage or fires. This can be monitored and identified prior to failure using infrared scanning or ultrasonic inspection methods. This information informs the maintenance team of electrical equipment that requires maintenance work, ideally scheduled to be executed during a planned electrical outage.

Electrical transformers require testing of the internal cooling oil. This would usually be scheduled to be conducted annually, using a trained technician to collect the oil samples. The oil samples are typically sent to an external qualified testing facility to perform specialized analysis. The analysis will indicate if there is contamination in the oil, and what may be causing that contamination. A tribology specialist will then identify corrective actions such as further inspection requirements or remedial actions such as filtering the oil to remove contamination, draining and replacing the oil, repairing internal components, or identifying if the transformer has degraded to the point of needing replacement.

An industrial facility of the scope and size of a uranium processing facility will have a significant amount of electrical cabling, supplying power from the switch room to the production equipment. These cables require protection from physical damage such as contact with moving equipment, but also protection from becoming overheated, which could happen if a large amount of dust or production material accumulates on the cable tray in which they are installed. Cable trays need to be kept clean and not be overloaded from their original design capacity. The trays are designed in a way to allow enough space between each cable for cooling purposes. Where practical, it is good practice to install cable trays in a vertical orientation so that material cannot build up on the cables. In an ageing facility, there may be times when obsolete equipment is replaced with new equipment. In this case the obsolete and unused electrical cables ought to be removed and disposed of to prevent overloading of the electrical cable tray as new cables are added. Removing obsolete cables will also limit confusion regarding which cables are still functional when performing condition inspections or during required maintenance work.

Backup power generation systems require maintenance to ensure they will operate when needed. When developing maintenance programmes, this critical equipment may incorrectly be overlooked, as it is not part of the electrical grid supplying constant power to the facility. A maintenance plan for backup power would typically include periodically testing the generators to verify that they function as intended. Large diesel generators are required to be manually turned, or 'barred' over, so that the internal bearings and mechanical components remain correctly lubricated and avoid becoming pitted by sitting in one position for an extended period. Lubrication oil and cooling systems need to be tested and analysed periodically to identify when replacement or filtering is required. This maintenance is to be performed even when the generator is not being used as moisture and contaminates can build up over time. An internal inspection of the mechanical components is also carried out by a trained technician on an interval indicated by the manufacturer. This would typically occur after a certain number of operating hours.

6.13. TAILINGS MANAGEMENT AND WATER TREATMENT

Effective management of tailings management facilities and water treatment facilities is essential to avoid production disruptions and for the protection of environment and the public. Tailings facilities are identified as one of the most significant operational and safety hazards in uranium mining and processing facilities, with potential long term serious impacts if not properly managed. As a waste stream from uranium processing, tailings retain radioactivity from the ore. While most of the uranium is removed during the processing of yellow cake product, radioactive decay products remain in the tailings. As such they may be a significant source of radiation exposure for several thousands of years, depending on the residual radioisotopes. Other metals in the ore and chemicals used in ore processing (e.g. arsenic, selenium, amongst others) are also typically present in tailings [85-87]. It is therefore vitally important to ensure geochemical stability of the contaminants contained in a tailings management facility as well as the geotechnical stability of tailing storage structures.

Tailings facilities may be engineered as above ground tailings dams or engineered in-pit tailings management facilities. Infrastructure for the operation of tailings management facilities will be required throughout the life of the mine, including during the closure and decommissioning phase. Critical assets associated with tailings facilities includes pumps, piping, infrastructure to support pumps and piping, electrical infrastructure and instrumentation control systems.

Directly associated with tailings management is water management, as tailings are often required to be dewatered to ensure geochemical and geotechnical stability. Water management in uranium mining and processing facilities also often includes the control and treatment of water either supplied to the site or discharged from the site [85]. As such, an effective water management strategy needs to be developed prior to the start of mining and processing of uranium. A water management strategy will be implemented at the start of mining and processing and be adjusted as required during operation of the mine and processing facility.

Globally, there are numerous types of water treatment facilities in operation in uranium mines and processing facilities. This includes, but is not limited to, reverse osmosis, nanofiltration, conventional lime neutralization water treatment plants, brine concentrators and evaporators. Although these water treatment facilities are unique in design and operation, they use common equipment such as process vessels, storage tanks, pumps, piping, support infrastructure, electrical infrastructure and instrumentation control systems. Maintenance strategies for both tailings and water treatment facilities can be set up in the same manner as uranium processing equipment.

At the time of mine closure and decommissioning it is likely that there will be sustained operation of assets associated with tailings and water management. Therefore, strategies for the lifetime of the mine need to consider the maintenance activities for tailings management and water treatment facilities in the long term, as these facilities may remain in operation, and as such require maintenance, for many decades post-closure.

Uranium mining companies are expected to demonstrate, through risk assessment over the lifetime of the facility, that the mine tailings and wastewaters during operation and post-closure will not adversely affect the environment or impact public safety [85]. As such, an effective maintenance strategy for these facilities is required to ensure these conditions are met.

6.14. SITE AND SUPPORTING FACILTIES AND SERVICES

The following subsections describe the types of infrastructure and services that are typically required to support a uranium mine and processing facility. In some cases, infrastructure may also service the public (e.g. roads) or nearby communities (e.g. electricity). As such, clearly established boundaries need to be in place in regard to infrastructure ownership, to determine accountability for maintenance of assets (i.e. by the owner/operator of the uranium production facility or a state owned enterprise), amongst other matters.

In some cases, the maintenance of specialist infrastructure such as bridges, rail lines, wharves and power stations may be outsourced to specialist contractors. It is important in these cases that the maintenance plans are still set up and tracked through the CMMS to ensure visibility of the asset condition and maintenance, as well as to receive the other associated benefits with following a structured work management programme.

6.14.1. Road

Vehicle access is normally facilitated through surface roads. Roads can be used to transport major and minor consumables, equipment, spares and personnel. Roads are to be well maintained to ensure the operation interconnects with national roads and other infrastructure, such as the closest port, airport or rail hubs.

6.14.2. Personnel transport

At some operations the mine provides transportation for personnel with a bus fleet. Vehicle fleet maintenance may be performed at on onsite facility workshop by qualified mechanics or at an offsite commercial service facility. Vehicle fleet maintenance is guided by the original manufacture service requirements and needs to be carefully scheduled and align with consideration of the shift structures and operating requirements. Replacement of ageing fleet will generally be funded via the operation's sustaining capital provision at optimal replacement intervals.

At fly-in-fly-out operations, personnel transport is via air transportation, in most cases through an operate and maintain contract with a commercial supplier. Additional considerations for maintenance of an airstrip and its associated infrastructure are also required, which may include facilities for aviation gas infrastructure for fuelling aircraft.

6.14.3. Rail

Rail transportation may provide an economical option for an operation to link with major freight centres through the national rail system. Direct rail access to a port is ideal for transport of bulk material. Further, rail transportation is deemed a relatively safe transportation option.

Major consumables such as sulphuric acid, diesel, ammonia, ammonium nitrate and containerized uranium product can be safely and economically transported via rail. It is beneficial to have separate offloading sidings for the mentioned major consumables as well as sidings to the various storage facilities for offloading of larger volume spares.

The railway and rail fleet are often state owned, operated and maintained. However, sections of the rail within the mine lease may be the responsibility of owner/operator to manage. Prior to construction of new railway tracks between the operation and national rail lines it needs to be ensured that the track gauge (i.e. distance between the rails) is compatible with the national lines.

6.14.4. Port

When available, a seaport can service the operation via a rail link or national road. Often the port is owned, operated and maintained by a state owned enterprise. A port will typically be used to import major consumables and to export containerized uranium product. A leased or owned facility in the port can provide consumable storage tanks, transfer facilities for offloading of a vessel and (if applicable) a rail loading facility to transfer a consumable to rail tankers.

6.14.5. Fuel

Fuel can be transported to the operation by either rail tankers or trucks. Suppliers are normally responsible for the supply and storage facilities on site, including maintenance activities. If operationally owned and operated, the maintenance regimes are normally in accordance with national and local regulations or aligned to industry standards.

6.14.6. Water

Water supply (and discharge) for industrial use and potable water consumption are important considerations for a uranium mine and processing facility to achieve operational efficiency and for the protection of human health and the environment.

A holistic water balance is required to determine the operational demand volumes and quality of water, manage groundwater, tailings facilities water, human consumption, dust suppression, rain event runoff, operational requirements and evaporation. A water balance model is typically developed during the environmental impact assessment phase of a uranium mine and processing facility as part of project application and licencing. As such, the operation will have detailed information about the water infrastructure and assets in the early phase of the project.

Water supply is often contracted with the local water utility via a water network consisting of storage reservoirs, pump and pipe systems. The local water utility will be responsible for maintenance of such infrastructure. It is therefore the responsibility of the owner/operator of a uranium mine and processing facility to ensure that onsite water infrastructure and assets are well maintained for operational requirements (e.g. water demand) and to ensure regulatory or licence conditions are met (e.g. environmental and health).

6.14.7. Power

An assessment of electrical requirements and availability is a fundamental consideration for the construction and operation of a uranium mine and processing facility. For example, during operations the electrical demand may vary due to seasonal conditions, maintenance shutdowns and daily demand patterns.

Power supply infrastructure may be connected to the national power grid, supplied by the operation itself or a combination of utility power, supplemented with the operation's own electricity supply. Electrical infrastructure and equipment may include power stations, substations, transmission lines, transformers, switchgear and distribution systems, cables, etc. Many operations are moving to low carbon footprint power supply options such as wind or solar. Backup power generation might be required for critical asset loads such as information technology infrastructure, security systems, emergency lighting, ventilation, etc.

Maintenance of electrical infrastructure is guided by national industry standards, or in some cases obligations are established in regulations, and is only to be performed by qualified and experienced high voltage and/or low voltage electricians. Safe work procedures for maintenance of electrical equipment requires the isolation of electrical energy sources with potential to interrupt power supply to critical assets. As such, electrical maintenance needs to be carefully scheduled and align with consideration of operational and production requirements. Information on maintenance of typical onsite electrical infrastructure is provided in Section 6.12.

6.14.8. Workshops and offices

Workshops and office structures need to be purposefully designed for the environmental conditions and function. The climate and structural aspects are key considerations in the design, construction, power consumption and maintenance efficiencies for this infrastructure. Section 2 describes requirements for safe and effective maintenance of occupied buildings, which includes both temporary or permanently occupied workshops and offices.

7. CARE AND MAINTENANCE

7.1. PLANNING FOR CARE AND MAINTENANCE

Uranium mines that are in care and maintenance are in a state typically characterized by the cessation of mining and/or processing operations and shutdown of facilities, which is generally described as temporary closure. A uranium mining operation may remain in care in maintenance for several years in anticipation of recommencing operations. In some cases, the operation may not restart after care and maintenance and will transition into permanent closure (see Section 8).

An organization may consider placing a mine or processing facility into care and maintenance when the operation, at a certain point in time, is unable to produce or when it is no longer economical to produce. Mines or processing facilities that are put into care and maintenance need to do so in accordance with the relevant guidance and licences issued by the regulatory body. The owner/operator needs to prepare a comprehensive care and maintenance plan that demonstrates that the facility is in a safe state and that workers, the public and the environment remain protected.

The duration of time of care and maintenance is often related to the price of uranium within the market. Other factors may include a change in government policy, a change in stakeholder acceptance, an unplanned failure of equipment requiring a prolonged period of downtime or environmental factors such as flooding. Judgement is required to understand the expected length of time the operation will remain in care and maintenance as the duration is an important consideration for developing a maintenance strategy. Short term care and maintenance would typically be less than three months (often while an issue is resolved) and long term care and maintenance would be greater than three months. Short term care and maintenance can be differentiated from partial care and maintenance. Partial refers to a specific area of the processing facility being taken offline and placed into care and maintenance while the remaining processing facility continues to operate. Additionally, the potential to extend the duration of care and maintenance is also assessed to ensure the correct care and maintenance processes are implemented.

While in care in maintenance, the production of uranium is stopped but the site is managed to ensure it remains in a safe and stable condition and be ready to restart. At this time the organization's operational plans will be modified for a state of care and maintenance, requiring the preparation of assets for this state, and revision of the overall maintenance strategy. Presumably, the organization has used valid economic assumptions and models in reaching a decision to suspend the operation of the assets and that sufficient funds are allocated for the care and maintenance activities. Generally, the maintenance strategy during care and maintenance will aim to preserve assets and equipment in a way that is aligned with the organization's objectives for future startup. However, a comprehensive assessment of the mine and processing facility infrastructure may be undertaken to review whether all assets need to be preserved, or alternatively replaced, at the time when the operation will be recommissioned. Original equipment manufacturers can provide meaningful inputs with regards to the preservation of the assets for both assessments and development of asset management strategies during care and maintenance.

This section provides general considerations and asset preservation strategies for facilities moving into a state of care and maintenance. This section also details the key considerations and strategies to be considered when restarting a mine or processing facility following a period of care and maintenance.

7.2. ROLES AND RESPONSIBILITIES

The accountability for managing assets during the different phases of care and maintenance may move between roles. For example, the operations manager or the maintenance manager may be accountable for decontamination and decommissioning of assets that are required to be shutdown. Once the decommissioning is complete, the accountability may be transferred to a 'care and maintenance manager'. On recommencement of operations, the accountability for recommissioning assets may again be transferred to an incoming maintenance or operations manager.

During all stages of the care and maintenance period, specific roles need to be allocated and responsibilities defined, for the maintenance of critical system such as fire systems, site security, safety systems, environmental monitoring and management, and any other equipment required for compliance obligations or in accordance with statutory requirements.

While it is beneficial to have the same personnel who were involved in the decommissioning stage on the site during recommissioning, this is not always feasible. As such, it is important that a detailed decommissioning and recommissioning plan is developed and records of decommissioning, care and maintenance and recommissioning activities are maintained.

7.3. RECORD KEEPING

Records of activities associated with care and maintenance need to be documented and be maintained in accordance with the organization's information management and information security management guidelines (see Section 5.5). The following aspects are key considerations related to asset management records during care and maintenance:

— Electronic documentation: Electronic documentation needs to be stored in the relevant server (e.g. corporate server located offsite or local server) that receives technical servicing and support.
— Hardcopy records: Engineering drawings, process control schematics and other hardcopy records need to be stored or archived in an appropriate central storage facility.
— Asset CMMS data: CMMS data is to be archived for re-activation at the time of facility recommissioning. However, if the CMMS is to remain active during care and maintenance, it needs be used to schedule, monitor and record any maintenance activities for continuity of data management.
— Warehouse and inventory data: are to be used to determine the current levels of stock, held on the site and off the site, including the locations of assets, reorder points and the maintenance programme for the equipment while in storage.
— Maintenance work orders: individual work orders are to be created for all activities associated with preparing the assets for care and maintenance. The functional location or asset numbers are to be used when creating the work order (see Section 4.1). Where required, the work orders need to be updated with accurate and relevant information to assist the recommissioning team for restarting operations at the end of the care and maintenance period.

7.4. SECURITY

Security of the site and of assets is important for uranium mines and processing facilities in care and maintenance, particularly in the case where nuclear material and radiation hazards remain. Valuable assets may become a target of theft during care and maintenance. In the worst case, there are examples of fatalities in mining industries occurring during the attempted theft of copper wire on live electrical circuits.

Facility assets need to be secured against criminal or intentional unauthorized acts, and the broader operational site also needs to be secured to protect the public (e.g. prevent public access to mine working areas). In full care and maintenance, the facility assets may be secured by security fencing, and either a permanent security presence on site or a regime of random site security checks. In partial care and maintenance, the decommissioned areas of uranium mine and processing facility will ideally be separated from the workforce, limiting access for only essential activities (i.e. area monitoring, inspection, maintenance). Physical barriers are most effective or if appropriate administrative controls (e.g. signage and training) can be used to declare the decommissioned areas as restricted access areas.

7.5. ENVIRONMENTAL FACTORS

The following environmental conditions need to be considered in the preservation of assets whilst under care and maintenance.

7.5.1. Fire

Avoiding fire related risks and fire damage is essential for not only asset health, but also for protection of people and the environment surrounding the operation. The following subsections describe fire management considerations.

7.5.1.1. Flammable substances

The storage of flammable substances and hydrocarbons is to be minimized and such substances need to be protected from lightning strikes and other potential ignition sources. Vegetation needs to be regularly cut back or removed during the care and maintenance period, providing appropriate clearance around any buildings or assets.

7.5.1.2. Fire Protection

Existing fire alarms, fire suppression systems and firefighting equipment, for both fixed and mobile assets, need to be available and maintained in a ready-to-use state, throughout the care and maintenance period. Any exception to this will require risk assessment with alternative control measures in place. In most cases, the management and maintenance of any fire alarms, fire suppression equipment and firefighting equipment will be subject to statutory or regulatory requirements which need to be fulfilled. These statutory requirements become the minimum standard for maintaining fire protection equipment.

7.5.2. Floods

During care and maintenance, management strategies and emergency response plans need to be in place to deal with the possibility of flooding. The preservation of assets and the protection of the environment and public are expected to be considered when planning for flood mitigation, water discharges from site or runoff, flood detection, flood response plans and repair of the site after flooding events. Additionally, a programme of inspection and maintenance of flood mitigation infrastructure needs to be in place for key assets such as water storage facilities, storm drains, sewers, ditches and other relevant equipment, such as weirs, water monitoring stations, pumps and pipelines.

7.5.3. Freezing weather

When freezing weather conditions are expected, steps are expected to be taken for the preservation of assets to prevent liquids from freezing in pipes, tanks and sumps, for example in water pipes, process tanks and fuel lines. Additionally, high density polyethylene pipelines are not to be taut, allowing for thermal contraction associated with temperature extremes. In regions where snow or hail is expected, an assessment of structures to identify assets that cannot support snow load or hail can indicate if the asset needs to be covered, removed or reinforced.

7.5.4. High winds

The operation needs to be prepared for high winds, especially areas prone to hurricanes or cyclones. Loose materials need to be consolidated and either secured or disposed. Flashing, guttering or tinwork on any structures or buildings needs to be periodically inspected and maintained to ensure it remains secured. Where available, assets are to be placed into lockdown storm positions (e.g. cranes in service positions, portable warehouses tied down to concrete anchors, clamped conveyors, etc.) and remain locked down during the care and maintenance period.

7.6. SPARES AND WAREHOUSED GOODS

If the operation is in partial care and maintenance, spares and warehoused goods will continue to be managed by a dedicated resource, typically the procurement team. The appropriate storage and maintenance practices will continue as normal, for example batteries need to be tested periodically and gearboxes and motors turned.

For operations in full care and maintenance, a full inventory stock take is expected to be performed. Where practical, all spares and warehoused goods need to be consolidated and centralized, ideally stored in a secure facility. As a minimum the spares and goods need to be protected by security fencing, in conjunction with either a permanent security presence on site, a regime of random site security checks or a reliable ‘back to base’ alarm system installed.

The following considerations need to be taken in to account:

— Environmental conditions: Location and weather conditions are to be considered when deciding on the storage requirements of spares and warehoused goods. For example, salt spray (i.e. at coastal locations) or humid weather conditions can be a source of corrosion and degradation of equipment.
— Shelf life: Spares holdings are to be reviewed for the presence of perishable goods or items with limited shelf life, and a decision made on whether to dispose of or sell these items.
— Offsite spares holdings: Spares held on the operation’s behalf by suppliers are to be sold or consolidated.
— Offsite repairs: Rotables and repairable equipment that requires repair offsite are to be assessed to determine whether further repairs need to be completed or halted, and whether the component remains in storage at the warehouse or disposed of.

— Consignment spares: Spares held on consignment from a vendor are to be returned to the vendor.
— Vermin and insect infestation: Measures are expected to be in place to prevent animal or insect infestations, such as mice or termites.

7.6.1. Hydrocarbons

Ambient temperatures and methods of storage will impact the long term storage of hydrocarbons, especially higher volatility fuel products as they tend to oxidize quickly.

Any hydrocarbons stored within drum containers need to be clearly labelled with the contents and date of receipt. The drums need to be kept undercover and stored horizontally. Any hydrocarbon within storage tanks, totes, pods, cubes and other similar vessels, need to be fitted with an appropriate desiccant breather.

As diesel fuel ages, fine sediment and gum forms. This results from the reaction of oxygen with parts of the fuel. The breakdown of diesel fuel can be accelerated by:

— Contact with copper, zinc or metals alloyed with either;
— High ambient temperatures (e.g. >30°C);
— Water (promotes bacterial or fungal growth);
— Dirt with trace elements that may react with the fuel (i.e. copper or zinc).

Fuel life can be prolonged by removing or controlling the conditions which accelerate fuel breakdown, by implementing the following measures:

— Where fuel may be in contact with copper, zinc or associated alloys, a metal deactivator additive can be used.
— Where possible, protect fuel storage tanks from direct sunlight to reduce temperatures.
— A biocide can be added to the fuel storage tank to prevent bacterial or fungal growth.
— An antioxidant additive can be used to prevent diesel oxidization and reduce the formation of sediment and gum.

7.6.2. Tyres

Tyres are generally treated as assets within the mining industry, with inspection and repair strategies in place to extend their lifespan. The storage and preservation of tyres will require consideration of the ownership of these assets in addition how long until they will be needed for use. The following are considerations for tyre storage:

— General Storage: Tyres are to be kept clean and where possible stored out of direct sunlight.
— Consignment tyres: Tyres held on consignment from a vendor are to be sent back to the supplier if they are not going to be needed for more than 12 weeks (i.e. long term care and maintenance).
— Tyres owned by the operation: Long term storage management aims to maximize the tyre life. The maximum long term storage period for a tyre is typically 5 years. In some cases, an alternate option to long term storage would be to either ship them to another operation (where available) or to sell them before their use by date expires.

7.7. WORKSHOP FACILITIES

Workshop facilities need to be thoroughly cleaned and refuse removed prior to care and maintenance and scheduled periodically during. A programme of maintenance needs to be developed for workshops, including structural inspections and planning for corrosion protection to be monitored during care and maintenance (see Section 6.2 for guidance on steel structures). In the case that workshop facilities will still be used during care and maintenance, Section 2 describes requirements for safe and effective maintenance of occupied buildings, which includes both temporary and permanently occupied workshop facilities.

All tooling and equipment is to be identified, catalogued and documented within the CMMS or as hard copy documentation for archiving. An assessment of workshop facility, machinery tools and consumables needs to be performed to understand if any equipment will be utilized during care and maintenance. Only the tools and consumables that are not required during care and maintenance are to be decommissioned and put into storage. Additionally, workshop equipment on consignment or being rented, that is not required during care and maintenance, is to be returned to the respective vendors.

The following subsections outline considerations for storage of workshop installations and tools.

7.7.1. Hand tools

The following considerations are relevant to hand tool storage:

- Metal hand tools are to be coated with a suitable rust preventor and stored in a secure, dry location.
- Electric hand tools are to be cleaned and wrapped with plastic (or placed in plastic bags) containing a suitable desiccant and stored in a secure, dry location.
- Air tools are to be prepared by attaching in-line lubricators, increasing the lubrication dose and running the tools (with dry compressed air) until there is no more visible oil coming out of the discharge holes. Excess oil is to be cleaned off the tool surface. Air tools are to be wrapped with plastic (or placed in plastic bags) containing a suitable desiccant, and stored in a secure, dry location.

7.7.2. Machine tools

Equipment used in metal working (e.g. shaping, cutting, machining) or for other rigid materials, such as a grinders or lathes, are regarded as machine tools. The following considerations are relevant for larger machine tool storage:

- Tools are to be cleaned.
- Any sharp cutting tools are to be removed.
- All bare metal surfaces are to be coated with a corrosion inhibitor.
- A desiccant or a volatile corrosion inhibitor emitter is to be placed inside any electrical cabinets.
- A volatile corrosion inhibitor is to be added into any oil filled gear or hydraulic compartments.
- Tools are to be covered with a waterproof material or (preferably) wrapped with plastic.

7.7.3. Cranes

Prior to care and maintenance, workshop cranes need to be moved into their service positions. Corrosion protection needs to be applied to all gearboxes, surfaces, switches and other susceptible components. Additionally, all lifting equipment associated with the crane (i.e. chains, slings, hooks, shackles, etc.) need to be cleaned, metal items coated with a suitable rust preventative, and where feasible, stored in a secure, dry location.

Crane inspections are usually mandated by regulations for registered equipment, with compliance certificates for inspection issued for define periods. During care and maintenance, scheduled inspections are expected to occur as normal.

As part of the equipment shutdown steps to prepare workshop cranes for care and maintenance, cranes need to be isolated from electricity and the control panels locked out to prevent use.

7.7.4. Maintenance consumables

Prior to care and maintenance, maintenance consumables such as, bolts, hydraulic fittings, welding consumables, fuses, etc., ideally will be identified, catalogued and consolidated in a secure centralized location. These components are to be packaged and stored, ready for recommissioning activities, or otherwise available for essential activities during care and maintenance.

7.7.5. Compressed gasses

Full gas bottles are expected to be stored in a secure warehouse facility, following the manufacturers requirements for correct storage, or alternately returned to the supplier for credit. Partially full gas bottles need to be sent back to the supplier. Compressed gas systems need to be drained of any moisture, and in the case of combustible or corrosive gasses the pipeline purged with either an inert gas (i.e. nitrogen) or dry compressed air.

7.8. CORROSION

Corrosion is one of the most prevalent and commonly occurring forms of degradation of assets in care and maintenance. The common types of corrosion are:

- Generalised corrosion: Presents as a well distributed attack on large areas of metal and can be caused by improper or compromised surface protection, coupled with environmental factors such as humidity or salt water.
- Pitting corrosion: Localized and deep penetration of the metal surface. This usually occurs due to protective coating failure associated with not cleaning structures or components thoroughly, or localized galvanic reactions.
- Galvanic corrosion: Localized corrosion due to electrochemical reactions propagated by the proximity of at least two dissimilar metals in a conductive environment.
- Microbial influenced corrosion: Caused by the presence of microbiological agents within specific environmental surroundings (e.g. temperature, pH, water quality). This is particularly relevant to heating, ventilation and air-conditioning pipework.
- Corrosion under insulation: Normally results from compromised or incorrect insulation protection leading to moisture penetrating insulation and remaining present at the metal surface.

It is likely that most types of assets (e.g. buildings, tanks, vehicles) will require some level of corrosion protection and ongoing monitoring during care and maintenance. A key activity is the preparation of the required assets for corrosion control, and the ongoing work to control corrosion during the period of care and maintenance.

Assets susceptible to corrosion are expected to be identified, and a plan for corrosion protection developed prior to, and implemented during, care and maintenance. An effective plan will detail the type of corrosion, corrosion protection measures, and monitoring requirements with consideration of the metal type, the function of the metal, the duration of care and maintenance and environmental conditions.

Removal of any chemicals, dirt or processing materials from contact with metal surfaces, and isolation from the presence of oxygen and/or water, will mostly halt further corrosion. Where possible, steps are to be taken to prevent or minimize potential galvanic corrosion. Where it is not possible to separate the dissimilar metals and/or neutralize the conductive environment, cathodic protection may be considered as a solution to reduce corrosion rates (e.g. a sacrificial anode may need to be installed).

Internal corrosion can threaten the preservation of assets as it is challenging to detect. Some of the methods of dealing with internal corrosion are:

- Volatile corrosion inhibitor (VCI): VCI is also known as vapour phase inhibitor and vapour phase corrosion inhibitor. VCIs are compounds transported, in a closed environment, to potential sites of corrosion by volatilization from a source. In boilers, volatile basic compounds, such as morpholine or hydrazine, are transported with steam to prevent corrosion in condenser tubes by neutralizing acidic carbon dioxide or by shifting the surface pH towards less acidic and corrosive values. In closed vapour spaces, such as shipping containers, volatile solids such as salts of dicyclohexylamine, cyclohexylamine and hexamethylenediamine are used. On contact with the metal surface, the vapour of these salts condenses and is hydrolysed by any moisture to liberate protective ions. VCIs are available in the form of a number of commercially available products. They can be purchased as an oil additive that does not interfere with the lubricating properties of the oil in the compartment, as a solid 'emitter' that can be placed inside an electrical cabinet, or as impregnated wrapping and packaging material. Care needs to be taken in the selection of VCI's. It is good practice to use VCIs with no volatile organic compounds and they are not to be used in combination with most types of desiccants.
- Desiccant: Is a substance, such as calcium oxide or silica gel, which has a high affinity to water. Desiccants can be placed within enclosures or pipe systems to help keep the air dry which will eliminate or reduce condensation within the space.
- Completely fill compartments: Compartments can be filled with suitable oil, engine coolant, or in some cases demineralized water to prevent corrosion.
- Inert gasses: Purging of system with (a compatible) inert gas acts to displace oxygen. Once they have been purged and dried out, sealed tanks and pipe systems can be filled with an inert gas to protect against internal corrosion.

Section 7.8.1 provides a case study example of completely filling compartments and purging systems with inert gases, for corrosion control.

Surface corrosion is easier to detect than internal corrosion. All steel surfaces need to be completely encapsulated in an appropriate protective film such as paint, electroplating, grease, oil or polyvinyl chloride coating. Metals like stainless steel, copper and bronze are generally considered to be corrosion resistant; however, in certain environmental conditions even these metals will need to be protected.

7.8.1. Carbon steel pipework

For long term care and maintenance, the following controls are to be considered for the preservation of carbon steel pipework in an empty or full state.

When keeping carbon steel pipework in a full state, that is completely filled compartments, it is good practice that the liquid is circulated within the pipework. Liquid circulation can be done slowly, with a full system turnover occurring over several days being acceptable. To minimize costs, water can be used as the fill medium so long as the right parameters are met. If using water, the preference is for it to be demineralized, or if that is not possible, rainwater may be considered. Heavily mineralized water (e.g. bore water) would be the least preferred water based fill medium. The pH of the water needs to be raised to around 9.5 and an organic oxygen scavenger added to help remove any dissolved oxygen from the system. One of the advantages of keeping the pipe system full is that it may also be possible to start and run pumps under care and maintenance, depending on the fill medium.

When keeping carbon steel pipework in an empty state all attempts need to be made to dry the inside of the pipe system completely. Draining the pipework and then running dried compressed air through the system can aid drying of pipework. Where plant compress air is not available in sufficient capacity, fan forced atmospheric air can be used. A hygrometer can be used to check how dry the system is by measuring the amount of water vapour contained in the outlet air. Once the carbon steel pipework has been sufficiently dried, a VCI emitter or a desiccant can be added, and the system sealed to keep out any additional moisture. An alternative to compressed air might be to dry the system by flushing it with an inert gas (i.e. nitrogen). In addition to drying the system, this also has the effect of purging the pipework of oxygen. When the oxygen in the system reaches a level of $< 1\%$, then the system can be sealed and positive pressure maintained using regulated pressurized inert gas.

7.9. SHORT TERM AND LONG TERM CARE AND MAINTENANCE PREPARATIONS

Assets under care and maintenance tend to benefit from a structured regime of regular startup and running as it keeps components lubricated, and prevents oxidization and degradation of components. Where it is not feasible to conduct regular startup and running of a processing plant during care and maintenance, couplings need to be disconnected to allow rotation of individual motors, gearboxes, compressors and other rotating equipment.

A decision needs to be made regarding the viability of totally draining the processing plant and removing all process materials and final product. In general, the decision will be dependent on the expected duration of the care and maintenance period. As such, there ought to be a process in place for review of this decision during care and maintenance. For example, a time based trigger, or event based trigger such as the period of care and maintenance period being extended or other external factors including, inter alia, changes in regulatory requirements or the organization's objectives.

This section of the publication provides general guidance on asset management strategies for uranium mines and processing facilities in short term and long term care and maintenance. The short term strategies are typically for periods less than 12 weeks, while long term strategies are typically for periods of more than 12 weeks. This section provides information on areas of processing plants, as well as for heavy mobile equipment (HME), such as hydraulic digging units, haul trucks, road dressing equipment, mobile cranes, forklifts and any other ancillary diesel equipment.

Table 17 outlines the basic asset management strategies when putting assets of uranium processing facilities into care and maintenance.

Table 18 outlines the basic asset management strategies when putting HME assets of uranium mining and processing facilities into short term care and maintenance. Short term care and maintenance is not normally a significant issue, with HME being designed for exposure to environmental or weather conditions, allowing them to sit in the sun, the rain or snow for a few weeks.

Table 19 outlines the basic asset management strategies when putting HME assets of uranium mining and processing facilities into long term care and maintenance. As a minimum, all the short term preparatory activities also need to be completed for the long term care and maintenance.

TABLE 17. ASSET MANANAGEMENT STRATEGIES FOR URANIUM PROCESSING FACILITIES IN SHORT AND LONG TERM CARE AND MAINTENANCE

Area	Strategies for short term care and maintenance	Strategies for long term care and maintenance
Cleaning	The assets need to be cleaned and inspected for damage to any protective surface finishes such as paint or electroplating	
Sumps	Drain all sumps and leave the sump pumps energized to ensure the sumps don't overflow. If sumps cannot be left energized, the pump has to be moved to above the full level of the sump	
Pipework	Existing processing materials and product drained from the main circuits as well as any hazardous substances that might be in any secondary circuits. Flushing the system may be considered for short term care and maintenance, generally dependent upon what is running through the pipework. If the system contains any type of solids in suspension, anything that is corrosive to the pipe material or anything potentially unstable, it needs to be drained and flushed out	Pipework needs to be kept either completely full or completely empty. Once flushed, clean synthetic poly, stainless steel and copper pipework likely will not present any major corrosion issues. Poly piping may need to be protected from the sun. Carbon steel pipework will almost always be subject to potential corrosion issues
Valves	Coat any exposed valve threads and shafts with a suitable corrosion inhibitor	Apply corrosion protection as appropriate, for example through application of grease, a strippable film, coal tar epoxy or similar. Consideration may be given to removing expensive or difficult to access valves, blinding the pipes and storing the valves indoors
Open tanks	Where practical, tanks are to be drained. Lined tanks may be kept filled with water to protect the liner, with due consideration to the expected length of care and maintenance. Continue to monitor equipment filled with fluids at an appropriate frequency, to ensure they are kept full. Acid tanks to be considered as keeping full	Open tanks present remain exposed to the weather and are not likely to be able to be kept dry internally. The drain valve needs to remain open and installation of some form of cathodic protection considered. Another option might be to allow a thin surface rust to develop on the tank and to apply a phosphoric acid based rust converter that acts to neutralize the corrosive action and turn the surface rust into a protective coating
Closed tanks	Where practical, tanks are to be drained. Lined tanks may be kept filled with water to protect the liner, with due consideration to the expected length of care and maintenance. Continue to monitor equipment filled with fluids at an appropriate frequency, to ensure they are kept full	Closed tanks are to be drained, cleaned and dried then filled with an inert gas and (where able to be pressurized) maintained at a suitable positive pressure. Where unable to be pressurized, an appropriate VCI emitter can be installed
Centrifugal pumps	Run all product out of the pump and flush, lubricate the pump, and treat to inhibit corrosion	Remove any gland packing and spray a suitable corrosion inhibitor into the stuffing box. Apply a suitable corrosion inhibitor to any exposed steel shafts. Open the pump and fog internals with a suitable corrosion inhibitor
Reciprocating pumps	Run all product out of the pump and flush, lubricate the pump, and treat to inhibit corrosion	Remove any gland packing and spray a suitable corrosion inhibitor into the stuffing box. Apply a suitable corrosion inhibitor to any exposed steel shafts. Open the pump and fog internals with a suitable corrosion inhibitor

TABLE 17. ASSET MANANAGEMENT STRATEGIES FOR URANIUM PROCESSING FACILITIES IN SHORT AND LONG TERM CARE AND MAINTENANCE (cont.)

Area	Strategies for short term care and maintenance	Strategies for long term care and maintenance
Compressors	Drain all air receivers and leave the drain valves slightly open. Ensure an information tag is fitted to the compressor start switch indicating the drain valve is open	A suitable VCI additive can be added to the oil compartment or, in the case of a screw compressor, the receiver tank. The inlet flange or air cleaner can be removed, and the exposed internals can be fogged with a corrosion inhibitor and/or the suction and discharge valves can be closed, and the compression chambers purged and filled with nitrogen gas. Maintain the nitrogen pressure at around 5psi. Coat any exposed steel shafts with a corrosion inhibitor
Electric motors	Includes all electric motors such as those fitted to conveyors, pumps, screens, mills and crushers. Grease the bearings and fit caps to the grease nipples. If the motors have heaters fitted, consider energizing them	Lift carbon brushes from commutators or slip rings. Where sleeve type bearings are fitted, add a VCI to the lubrication system. Apply a suitable corrosion inhibitor to any exposed steel shafts. Motors over 40kW need to be set up so that they can be rotated during the care and maintenance phase
Transformers	Transformers are to be left energized to maintain temperature.	Ensure appropriate desiccant breathers are fitted.
Substations	Doors are to be kept closed and (preferably) locked. Air-conditioning to be left on, if possible	Consideration needs to be given to racking out the main isolation breakers, protecting them from dust and applying corrosion inhibitors
Electrical cables	Consideration needs to be given to the potential for vermin to damage electrical cables with the most common form of damage being to insulation caused by animals chewing through the cable insulation. The most common offenders are rodents and birds	
Electrical enclosures	Ensure electrical enclosures are sealed and locked appropriately	Ensure any door seals and inspection point seals are in good condition and place some form of desiccant and / or VCI emitter inside the enclosure
Variable voltage / frequency drives	Where to be de-energized, extract a list of the current operating parameters and settings and archive with the control systems data	
Process control systems	Ensure condensation is drained from any control system pipework. There needs to be a time based trigger in place for the capture and archiving of the control system and programmable logic controller configuration files in the event where the care and maintenance period is extended past short term care and maintenance	Upload electronic configuration files from the programmable logic controller and DCS, store files on the corporate server. All hard copy configuration files such as instrumentation calibration settings are to be archived and stored. Place a suitable desiccant and/or VCI emitter inside of any control system enclosures
Sensors and senders	Consider removing instrumentation components that would normally be in contact with the process materials. Sensors and senders need to be cleaned, protected and kept in local storage	
Batteries	Batteries are to be placed into storage in accordance with original equipment manufacturers' specifications. Every attempt needs to be made to leave the power on to battery chargers	
Gearboxes	Includes all gearboxes as fitted to for example conveyors, pumps, screens and crushers. Ensure there are desiccant breathers fitted to all gearboxes	Add a suitable VCI additive to the oil compartment, or alternatively, completely fill the compartment with oil. Apply a suitable corrosion inhibitor to any exposed steel shafts

TABLE 17. ASSET MANANAGEMENT STRATEGIES FOR URANIUM PROCESSING FACILITIES IN SHORT AND LONG TERM CARE AND MAINTENANCE (cont.)

Area	Strategies for short term care and maintenance	Strategies for long term care and maintenance
External gears and chain drives	Ensure all gear and chain surfaces are coated with the standard lubricant	
V-belts and pulleys	De-tension belts and pulleys	V-belts need to be removed and stored. The pulleys are best preserved if coated with a suitable corrosion inhibitor
Belt conveyors	Run all process material and product off conveyor system and clean up any spillage	Tensioning devices need to be removed
Screw conveyors	Run all product off conveyor system and clean up any spillage	Coat screw with corrosion inhibitor if necessary
Chutes transfer points, impact tables and rock boxes	Clean out completely and clean up any spillage	
Calciners and roasters	Run out all product and slowly cool the unit over at least 48 hours. Seal prior to commencement of care and maintenance period. Manual cleaning of any residual product	Contact OEM for specific information relevant to preparations for long term care and maintenance. Annual visual inspection of the refractory. Steps taken to avoid moisture ingress
Plate and vibrating feeders, screens	Run all process material and product off of the feeder and clean. Clean up any spillage	
Winders and hoists	Flow to the rope lubricators need to be increased and ropes need to be cycled enough times to ensure full lubricant coating of all active rope areas. Any non-lubricated rope has to be lubricated manually with standard rope lubricant. The cage or skip has to be fully raised and locked in place leaving enough weight on the ropes to prevent kinking or excessive movement	Where there is not going to be a regime of regular startup and running, secure the rope drum and jack or wind the brake pads off of the brake path or the brake shoes off of the brake drum
Mills, crushers, sizers, plate feeders	Run all process material and clean and clean up any spillage. Coat all exposed metal components with a suitable corrosion inhibitor	
Vibrating feeders, screens	No specific strategies	Where poly or rubber decks are being used, they need to be either removed and stored or protected from the weather
Heat exchangers	Flush and store wet, full of an appropriate liquid, or alternately flush and store dry. Wet storage: Consider the metal types inside the heat exchanger such that informed decisions are able to be made regarding the storage medium Dry storage with inert gas: Once dried, the exchanger can be flushed with an inert gas (such as nitrogen) and stored full of an inert gas with a suitable positive pressure Dry storage with normal atmosphere: Once dried, a suitable VCI emitter and/or desiccant can be put inside and the unit sealed	

TABLE 17. ASSET MANANAGEMENT STRATEGIES FOR URANIUM PROCESSING FACILITIES IN SHORT AND LONG TERM CARE AND MAINTENANCE (cont.)

Area	Strategies for short term care and maintenance	Strategies for long term care and maintenance
Steel structures	No specific strategies.	Need to be inspected for corrosion and for the integrity of the coating. Any damaged paintwork needs to be repaired or noted on the corrosion inspection form, to be monitored during care and maintenance
Hydraulic systems	Hydraulic cylinders: Exposed hydraulic pneumatic cylinder rods are to be coated to prevent corrosion Valve spools: Exposed valve spools need to be coated to prevent degradation Hydraulic hose ends and fittings: Plated hose ends and hydraulic fittings need to be coated with a suitable corrosion inhibitor Hydraulic hoses: Protect from direct sunlight Hydraulic systems: Drain off any water from the hydraulic tank and add a suitable VCI additive	

TABLE 18. ASSET MANAGEMENT STRATEGIES FOR URANIUM MINING HEAVY MOBILE EQUIPMENT IN SHORT TERM CARE AND MAINTENANCE

Area	Strategies for short term care and maintenance
Cleaning	Assets need to be cleaned down properly and inspected for damage to the painted surfaces
Roll away	Assets have to be protected from potential roll away by using wheel chocks, spoon drains, wind rows or similar
Batteries	Batteries are to be disconnected to minimize the chance of fire and to maximize battery life. A 12V battery will lose up to 1% of its charge per day if left idle, so it is advisable to install a trickle charger that will maintain around 13.1 volts at the battery or when set up in series 26.2 volts. Solar chargers work well in most parts of the world and require minimal maintenance
Fuel system	A diesel fuel biocide may be added to the fuel tank
Belly dumpers	Belly doors left open to prevent the body filling with rainwater or snow
Rear tippers	Truck bodies need to be left in the raised position to prevent the body filling with rainwater or snow. Ensure the ‘body up’ cables or pins are installed to prevent uncontrolled lowering of the body in the event of a hose or hydraulic failure
Side tippers	Truck bodies need to be left locked in the tipping position to prevent the body filling with rainwater or snow
Dozers, graders, excavators, shovels, forklifts, tyre handlers, front end loaders	Implements lowered to the ground
Forklifts	Coat any chains, ropes and sliding surfaces with a suitable corrosion inhibitor
Engine powered ancillary equipment such as mobile welders, mobile generators, mobile compressors	Maintain as per the guidance for the individual components such as tyres, compressors, generators, batteries and engines
Drill rigs	Masts need to be locked down in the tram position
Mobile cranes	All sheaves, pins, blocks, ropes, drum grooves, and miscellaneous hardware need to be greased. Exposed ring and pinion gear and bearing surfaces are to be coated with an appropriate brush or spray on gear/lubricant
Tyre handlers	Coat any sliding surfaces with grease or a suitable corrosion inhibitor

TABLE 19. ASSET MANAGEMENT STRATEGIES FOR URANIUM MINING HEAVY MOBILE EQUIPMENT IN LONG TERM CARE AND MAINTENANCE

Area	Strategies for long term care and maintenance
Surface corrosion	Assets need to be assessed for potential corrosion issues with the approach to corrosion control informed by environmental conditions and the quality of the existing coatings. It is important that assets are cleaned properly, particularly in difficult to access locations, notably inspection points. Assets need to be cleaned as dirt left behind will promote corrosion
Engines	Add a VCI additive to the engine oil. Coat the engine with a corrosion inhibitor. A properly applied coat of paint will protect the engine for up to two years. Change out the engine oil and filter if the engine has more than 50 hours operation
Fuel system	Consider draining the fuel injection pump and refilling it with a suitable liquid. Add an antioxidant and a VCI additive to the fuel tank
Drill rigs	Apply suitable corrosion protection to all steel surfaces. Drill rods and consumables need to be removed and stored in a secure location
Hydraulic systems	Hydraulic cylinders: Exposed hydraulic and pneumatic cylinder rods are to be coated to prevent corrosion Valve Spools: Exposed valve spools need to be coated appropriately to prevent degradation Hydraulic hose ends and fittings: Plated hose ends and hydraulic fittings need to be coated with a suitable corrosion inhibitor Hydraulic hoses: Protected from direct sunlight Hydraulic tank: Drain off any water from the hydraulic tank and add a suitable VCI additive
Electrical switches	Where appropriate, coat with waterproof silicon grease or similar
Infestation	Measures need to be undertaken to ensure vermin (e.g. rodents, birds or insects) do not infest the cabins
Electric (wheel) motors, generators/alternators	Where there is not going to be a regime of regular startup and running, lift carbon brushes from commutators or slip rings
Electric drive axle boxes	Ensure the door seals are competent, fit a desiccant breather and place some form of desiccant or VCI emitter inside the axle box enclosure
Electrical enclosures	Ensure any door seals are competent and place some form of desiccant or VCI emitter inside the enclosure
Wheels and tyres	Tyres are to be kept clean and where possible stored out of direct sunlight. As a minimum, place rubber tyres on stands to elevate the tyres off of the ground. Check the expiry date on all HME tyres and wheel rims and detail this information in the recommissioning plan
Weather	Where practical, park HME under cover. If that is not possible, black out the windows to prevent sun damage to the interior of the cabin
Internal corrosion all compartments	Either add a VCI additive to all oil or fuel filled compartments. Alternately if there is no regime of regular running of the asset, completely fill the compartments with oil or in the case of the fuel tank with fuel. Either completely seal or fit desiccant breathers to compartments
Mobile cranes	Where practical, lower the boom onto a suitable stand. All bearings, couplings, reducers and limit-switch gear cases on cranes need to be filled with the required lubricants. Rollers, roller path, kingpins, collector-ring assemblies, boom-hinge pins, metal surfaces of friction clutches, brakes, bearing surfaces, moving contacts and all other unpainted parts subject to corrosion need to be greased using waterproof grease or coated with a suitable corrosion inhibitor. All sheaves, pins, blocks, cables, drum grooves, and miscellaneous hardware need to be greased. The heavy grease in bull gears ought to be retained

7.10. SCHEDULING OF TASKS

Preservation of assets during care and maintenance is best managed using the same work management processes as used during normal operations. Work orders are to be raised for all inspections and work tasks, and schedules produced to drive the activities. Good practice involves work management records being archived for future use. All work orders need to be raised against the relevant functional location or asset number to enable review of how well the asset is being looked after and for the recommissioning team to review care and maintenance history.

7.11. RESTARTING OPERATIONS

If the owner/operator of a uranium mine and processing facility that has been in care and maintenance decides to restart operations, then a comprehensive assessment of the mine and processing facility infrastructure needs to be completed to ensure that the site is in a good physical and mechanical state and is safe to operate. A detailed and systematic recommissioning plan is to be developed to bring the mine and processing facilities online. This plan is ideally developed as part of the care and maintenance strategy (prior to shutting down), as asset preservation strategies determine the level of effort required to bring the facility back to full operations. This plan is developed with the inputs from the OEMs, as in most cases, recommissioning of assets closely resembles the OEM commissioning procedures.

7.11.1. Roles and responsibilities

On restart of operations, the accountability for recommissioning assets will need to be determined. Incoming maintenance and operations senior management will generally be appointed with accountability to oversee execution of the recommissioning plans. Execution of recommission plans is a complex set of tasks requiring technical and leadership oversight, training a largely new workforce and management of contractors. Therefore, it is good practice to appoint a project manager to lead the execution of the overall recommissioning plan. The project manager will be accompanied by a scheduler and a staff of detailed planners. Qualified engineers and equipment inspectors will be required to validate the state of the equipment and provide quality assurance. It is valuable to have people who were involved in the decommissioning phase available to participate in the recommissioning activities. As this is not always feasible, the record keeping of decommissioning activities are important for safe and effective recommissioning.

7.11.2. Record keeping

All recommissioning maintenance tasks need to be planned and executed using the organization’s same work management processes that were used prior to and during execution of care and maintenance. Recommissioning work orders are to be created as follow-up work orders, ideally linked in the CMMS to the work orders that were used to execute the care and maintenance tasks. This approach will make the process simpler for the planning team, so that they are able to revoke all the asset preservation activities in a systematic manner.

7.11.3. Recommissioning assets

The recommissioning plan will include the following general considerations for restart of assets:

- Spares parts inventory: Inventoried spares and stocking levels in the warehouse will need to be replenished. An assessment of needs is performed to determine if inventory and stocks need to be replenished to the same levels prior to care and maintenance, or alternate levels suitable to new planned operating conditions (e.g. higher or lower production targets). Restocking is typically as simple as reactivating the asset in the CMMS and issuing purchase orders to vendors. This is to be completed with sufficient time, prior to facility restart, to allow longer lead time items to arrive.
- Hydrocarbons, reagents, process chemicals: All fuels stored at site during care and maintenance ought to be tested to ensure they are still safe to be used. Fuel tanks may need to be drained and cleaned prior to being refilled. Acid needs to be tested and may need to be filtered. Process chemicals that were stored on site during care and maintenance require laboratory testing to ensure they maintained their intended composition prior to being used.
- Workshop facilities: Consumables and tools need to be reordered and made available for use and brought back to pre-care and maintenance levels, alternate levels suitable to new planned operating conditions.
- Rental equipment: An assessment to identify any rented equipment required for recommissioning assets or for normal full-scale operations ought to be completed in sufficient time prior to facility restart, and arrangements made with vendors to provide equipment to allow longer lead time items to arrive.

7.11.3.1. Assess for corrosion

Corrosion protection planning initiated at the start of care and maintenance ought to have protected assets as intended, however each asset needs to be inspected to determine current state prior to startup. Typically, skilled trades workers perform a basic visual inspection, and if an item or component appears to be of concern, further inspection and corrective action is determined by an engineer. If the care and maintenance activities were executed, observations of the state of the equipment were recorded during care and maintenance, and the corrosion protection strategies in place were followed, then there is minimal concern with identifying new corrosion issues. The interior of pipes, vessels, tanks, gear boxes, etc., may present a challenge. This equipment needs to be dutifully inspected using the same methods used in failure identification and preventive maintenance regime, such as thickness testing of a vessel from the exterior, internal visual inspections, and other advanced methods based on the criticality of the asset.

7.11.3.2. Asset specific considerations

For all equipment, clean off any preservation fluid or grease, as it may have accumulated dust and other contaminants, and lubricate as per the OEM specification. Strategies for asset recommissioning of specific assets of uranium processing facilities are provided in Table 20.

TABLE 20. STRATEGIES FOR RECOMMISSIONING ASSETS OF URANIUM PROCESSING FACILITIES AFTER CARE AND MAINTENANCE

Area	Recommissioning strategy consideration
Pipework	Completely flush any preservation fluids from the pipeline using a fluid that is compatible with the operational chemistry. This may simply be clean process water
Tanks (carbon steel)	Inspect, clean as needed
Pumps	Replace gland packing. Perform a baseline vibration measurement to identify any bearing issues
Compressors	Filter the oil using a kidney loop system to remove any contaminates, lubricate as per OEM specification. Perform a baseline vibration measurement to identify any bearing issues For electric Motors, re-fit the carbon brushes if lifted for care and maintenance, lubricate as per OEM specification. Perform a baseline vibration measurement to identify any bearing issues
Transformers	Take oil sample and perform analysis. Perform corrective actions identified as a result of the analysis
Substations	Lubricate as per OEM specification
Electrical cables	Perform visual inspections, checking for damage to cable insulation
Electrical enclosures	Return to original state, remove desiccant and/or VCI emitter from inside the enclosure
Digital control systems	Re-enable programmable logic controller and DCS, install software updates and patches as per OEM specification. Remove desiccant and/or VCI emitter from inside the enclosures
Sensors and senders	Where removed, re-install and calibrate all sensors and instruments
Gearboxes	Filter the oil using a kidney loop system to remove any contaminates, set oil level as per OEM specification. Perform a baseline vibration measurement to identify any bearing issues
Torque converters	Filter the oil using a kidney loop system to remove any contaminates, set oil levels as per OEM specification. Perform a baseline vibration measurement to identify any bearing issues
V-belts and pulleys	Re-install v-belts, checking for cracking and deterioration of the rubber. Validate alignment, and tension as per OEM specification
Belt conveyors	Check for cracking and deterioration of the rubber. Validate alignment, and tension as per OEM specification. Inspect all rolling elements, replace or repair as necessary
Calciners and roasters	Contact the OEM for specific information relevant to recommissioning. At a minimum a complete inspection of the refractory and fuel train is required.
Grinding mills, crushers and sizers	Filter the oil using a kidney loop system to remove any contaminates, set oil level as per OEM specification. Perform a baseline vibration of the drive system to identify any bearing issues
Vibrating feeders	Where poly or rubber decks are being used, they are fully inspected for weathering, and tested for durometer hardness measurement to be within manufacturer's specifications, otherwise they will likely crack upon initial startup
Screens	Where poly or rubber decks are being used, they are fully inspected for weathering, and tested for durometer hardness measurement to be within manufacturers specifications, otherwise they will likely crack upon initial startup
Hydraulic systems	Inspect the exterior of hydraulic hoses for weathering, if cracks are obvious, replace. Drain hydraulic systems of any water from the hydraulic tank, filter the hydraulic fluid using a kidney-loop filter system, replace system filters, top up hydraulic fluid to the OEM specification. Start system and inspect for any leaking seals, replace as required
Winders and hoists	Refer to manufacturer's initial commissioning instructions, inspect the brake pads or brake shoes. Test that brakes work as intended. Validate that regulatory requirements are met
Heat exchangers	To be flushed and leak tested. Any seals that continue to leak ought to be replaced. Clean exterior of any dust or debris
Steel structures	Inspected for corrosion and for the competence of the existing coating. Any damaged paintwork needs to be repaired. If the steel itself is degraded beyond safe loading conditions, replace as required

7.11.3.3. *Heavy mobile equipment considerations*

Restarting HME after short term care and maintenance is not normally a significant issue, as these types of assets are designed for environmental exposures and a range of weather conditions. However, the following are to be considered when restarting HME after short term care and maintenance:

- Batteries: Test batteries for load capacity, replace as required.
- Tyres: Check all tyre pressures, top up as required.
- General: Clean any preservation fluids from surfaces, lubricate in accordance with OEM specification.

Table 21 outlines asset management strategies for recommissioning HME assets after long term care and maintenance. As a minimum, all of the general and short term restart activities above apply.

TABLE 21. STRATEGIES FOR RECOMMISSIONING HEAVY MOBILE EQUIPMENT OF URANIUM MINING FACILITIES AFTER LONG TERM CARE AND MAINTENANCE

Area	Recommissioning Strategy Consideration
Engine	Check engine oil level, top up as necessary. Remove added corrosion protection fluids or preservation grease that were applied, lubricate in accordance with OEM specification. Where there is no regime of regular running of the asset during care and maintenance the following applies Fuel System: Drain the fuel injection pump, replace fuel filter, drain fuel tank of old fuel, refill system, and prime pump V-belts: Inspect v-belts for weathering and cracks, reinstall as per OEM direction Sealing: Remove any added seals that were placed on the intake and exhaust of the engine Cooling system: Sample coolant fluid, ensure its properties remain, replace if required Fluid levels: Return all fluid levels to OEM specification, take a sample of the engine oil, test for moisture content, replace or filter if required
Fluid level	Return all fluid levels to OEM specification. Take a sample of the engine oil, test for moisture content, replace or filter if required
Infestation	Perform detailed inspection for nests or signs of vermin and clean as appropriate. Typical areas to check include the vehicle cabin, fresh air filter housing, engine air filter, etc.
Frames and structures	Visually inspect all frames and structural components for corrosion and damage. If any is identified perform a structural engineering assessment to determine necessary corrective measures
Electrical (wheel) motors, generators and alternators	Where there was no regime of regular startup and running, replace carbon brushes on commutators or slip rings
Electric drive axle boxes	Inspect enclosures for damage (including rodent nests), and any subsequent damage to wires, repair as required
Electrical enclosures	Inspect enclosures for damage (including rodent nests), and any subsequent damage to wires, repair as required
Wheels and tyres	Inspect all rubber tyres for weathering and validate that they are within their useful lifespan, if not replace as required. Check the expiry date on all HME wheel rims. The rims may require a complete inspection prior to entering service
Drill rigs	Clean of any preservation fluid or grease, as it may have accumulated dust and other contaminants. Perform other general inspections required for all other HME components
Mobile cranes	Clean of any preservation fluid or grease, as it may have accumulated dust and other contaminants. Perform all other inspections required for all other HME components. Most jurisdictions have requirements for an engineering inspection at a certain frequency regardless of the hours of use between inspections. Consult national regulations regarding inspection of lifting devices and act accordingly. This may apply to ropes, hoists, slings, and associated rigging components

7.12. PROCESSING FACILITY STARTUP

The process to restart a uranium mill and processing facility is complex and highly dependent on how the process is designed and intended to function. However, the elements of a restart plan will share many of the following considerations:

— Return all mechanical assets from preservation state, and back to operational state as per asset specific directions (i.e. as described in Section 7.11).
— Remove isolating blanks from process piping and vessels.
— Test all electronic circuitry, power systems, control systems.
— Startup ventilation equipment and validate air flow measurements.
— Confirm all safety components are in place, PRVs, PSVs, machine guarding, alarm systems, and other process safety components and controls.
— Charge the compressed air system, blow down all water traps, replace inline air filters and desiccant type filters, verify clean air prior to activating any pneumatic equipment.
— Charge the plant water system, flush all lines, as rust and contaminants may have built up. Inspect for leaks, repair as required. Replace all inline filters and verify adequate flow rate of water systems. Typically, plant water is used to supply water to the gland seals on rotating pumps: flush gland water lines prior to reconnecting and validate that adequate flow rate at each gland seal is available, prior to starting any pumps.
— Pressure test all critical systems, pressure vessels, and piping systems such as steam, compressed gasses and ammonia as per regulatory standards.
— Test all safety stops on mechanical devices, verify that the control circuitry works as intended.
— Circulate water through the process system where appropriate. Inspect all pipe works, vessels, tanks, for leaks. Repair as required. Once operational, inspect and adjust gland seals on pumps as necessary.
— Begin to warm up calciners and dryers according to the OEM specified rate. This is critical so that refractory brick and the structural system expand at the correct rate.
— Validate flows of all reagents, including acid, if those are part of the process. Inspect for leaks and repair as required. Add flange covers to acid piping if they were removed. Good practice is to include a litmus style indicator on flange covers to draw attention to acid leaks.
— Lubricate all rotating assets as per OEM specifications. Take base line vibration testing on all assets as per the normal preventive maintenance strategy.
— Begin to add feed into the processing system as per the metallurgical or startup strategy.
— Perform frequent inspections in the initial hours and days as the plant is started up. Typically, equipment that has been dormant for a long period tends to have a high probability of initial failure, specifically on rotating assets.
— Initiate the normal preventive maintenance programme for the facility.

8. DECOMMISSIONING AND CLOSURE

The closure of a uranium mine and processing facility is a complex operational and regulatory task. A comprehensive closure plan detailing decontamination, demolition and rehabilitation needs to be developed by the facility owner/operator and approved by the regulatory body. Good practices for preparation of closure plans are described in Ref. [88]. During the planning an execution of a closure plan, considerable attention is required to determine the fate of assets and infrastructure and their maintenance strategies.

In this section, key considerations for asset management for facilities that are entering into closure, including the decommissioning phase, are provided.

8.1. MAINTENANCE REDUCTION

Assets typically do not need to be maintained to the same level when advancing towards closure. To minimize costs in the final stages of production, maintenance activities may be reduced while keeping the equipment operating. The cost benefits of reducing maintenance need to be weighed against the potential loss of production due to equipment failure. However, in the case of equipment failure that has potential for an undesired health, safety or environmental impact, maintenance activities are not to be reduced.

In making a decision to reduce maintenance, site specific conditions need to be considered, and this will depend on the condition of the asset and the level of risk associated with the change. An assessment of the current condition of the equipment is to be performed before making the decision to reduce major maintenance and refurbishment tasks. While reducing maintenance can also reduce costs, inherently it introduces additional risk of equipment failure. The organization needs alignment and clarity on the level of risk that is acceptable to the business. It is important to note that statutory or registered assets need to continue to be maintained in accordance with their required maintenance strategies.

General examples of equipment, and the types of planned maintenance that may be reduced, are outlined below in Table 22.

TABLE 22. EXAMPLE OF REDUCED MAINTENANCE AS A FACILITY NEARS CLOSURE

Area	Change to maintenance plans prior to closure
Electrical	**Change**
Thermal imaging of certain electrical equipment	Deactivate 3 months in advance
Tertiary and secondary crusher positions	Deactivate 3 months in advance
Tertiary and secondary 3, 6, 12 monthly inspections	Deactivate 1 year in advance
Yearly megohmmeter check inspections	Deactivate 3 months in advance
Change electric motor bearings, 3 to 6 monthly	Change 1 year in advance
Low risk equipment	Deactivate 1 year in advance
Inspections with frequency greater than 1 year	Deactivate 3 months in advance
Condition monitoring	**Change**
Motor, conveyor, thermal imaging	Change from monthly to 2 monthly at 6 months in advance
Electrical thermal imaging	Deactivate 3 months in advance
Thickness test of conveyor belts, 3 monthly	Deactivate 6 months in advance
Infrastructure	Change to maintenance in advance
Air-conditioner services	Change from 3 monthly to 6 monthly, 1 year in advance
Mechanical	**Change**
Inspections with frequency greater than 1 year	Deactivate 3 months in advance
Mill liner inspections	Deactivate 6 months in advance
Mill inspections	Change from 12 weekly to 24 weekly, 1 year in advance
Lime mill inspections	Deactivate 6 months in advance
Disassemble and reassemble ducting	Change from monthly to 2 monthly, 6 months, in advance

TABLE 22. EXAMPLE OF REDUCED MAINTENANCE AS A FACILITY NEARS CLOSURE (cont.)

Area	Change to maintenance plans prior to closure
Mechanical	**Change**
Disassemble and reassemble ducting	Change from monthly to 2 monthly, 6 months, in advance
Scrubber inspections	Change from monthly to 3 monthly, 6 months, in advance
Mill vent fan inspections	Deactivate 6 months in advance
Warehouse motor rotations	Deactivate 6 months in advance
Apron feeder inspections	Deactivate 1 year in advance
Major asset refurbishment	**Change**
CCD refurbishment	Stop 18 months in advance
Tanks internal and external inspections	Deactivate 3 months in advance

During any period of reduced inspection and maintenance, there is an increased reliance on production operators to identify unusual vibration, noise, or general poor condition of the asset which may indicate that the asset function is reducing. These issues then need to be escalated to the maintenance team to investigate and provide further assessment of the asset.

Leading up to cessation of operations, inventoried spares and stocking levels in the warehouse will need to be reduced. Typically, the spares reorder minimum can be changed to zero to minimize spares remaining within the warehouse once operations are completed. Decisions will need to be made about continuing to stock high value low usage spares during the final months of production. Alternate inventory strategies may include a payment to the supplier to reserve items, such that in the rare case that the spare is required it can be purchased in full and dispatched to site. Arrangements can also be made with suppliers to buy back remaining spares which are not required at the end of the operations.

In preparation for cessation of operations, a decontamination plan for the final production circuits, in particular the calciner and packing areas, is expected to be implemented. All products are expected to be recovered from the final circuit as part of final production operations. Section 8.2 provides additional information about decontamination and decommission of the broader mine and processing facilities.

8.2. DECONTAMINATION AND DECOMMISSIONING

The closure process requires plant to first be decontaminated and decommissioned. The decontamination process is the cleaning and flushing of all equipment, such that it is in a clean state until it is ready to be demolished. At the end of production, leaving uranium processing assets such as tanks, vessels and piping full of scale or processing liquids is to be avoided through a programme of decontamination works. Examples of decontamination works includes cleaning scale and sludge from tanks, flushing piping, clearing chutes, emptying media from filters and emptying unused processing substances from tanks and vessels. During the decontamination process, safe work procedures will require areas of the plant to be isolated from energy sources and process flows.

Decontamination activities form part of the broader decommissioning process. Decommissioning is expected to ensure that each area of the mine and processing facility can be left in a safe state long term and ensures equipment cannot be activated or pose a hazard to the environment or people. Examples of decommissioning assets includes removing drain valves from tanks to ensure any water can drain from them, removing sections of piping to prevent them from filling, and disconnecting power from equipment.

Where the decontamination and decommissioning steps are completed systematically, this establishes safe work conditions for the demolition team to remove and pull down assets with reduced risk of environmental contamination or injury to persons. A key aspect in demolition of facilities and assets is an understanding of structural integrity and degradation mechanisms. Ideally the plant will be demolished as soon as practicable after decommissioning has been completed to minimize plant degradation. This approach assists with retention of knowledge of the plant in preparation for, and while executing, demolition activities.

A high level order for asset management in the closure phase of a uranium mine and processing facility is as follows:

(1) Planning;
(2) Shutdown, flushing and isolation (for decontamination);
(3) Decontamination;
(4) De-energization and isolation (for demolition);
(5) Demolition.

The following subsections describe these steps in more detail, assuming that a period of care and maintenance will not be required. More information on care and maintenance can be found in Section 7.

8.2.1. Planning

Planning for decontamination and decommissioning can be completed through the development of work packages. Areas of the mine and processing facility can be broken into decommissioning zones, that group systems and plant processes so they may be logically shut down, decontaminated, de-energized and isolated. Each of these zones will have a decommissioning work package that details the work required. Decommissioning work packages document the status of the equipment and contain detailed information including:

- A check sheet with authorization by personnel responsible and accountable for executing the works and any special notes (e.g. items to be removed intact from the demolition area).
- Piping and instrumentation diagrams that identify:
  - Cleaning levels required for equipment, piping and in-line items;
  - Battery limits of equipment and piping systems for which a gas clearance certificate (e.g. anhydrous ammonia) is to be provided and location of sampling points;
  - Services isolation and tie-off points.
- Single line diagrams that identify disconnection points for electrical feeders in substations.
- General arrangement drawings identifying locations for:
  - Radiation clearance surveying;
  - Electrical cable tray disconnections at the demolition boundary;
  - Local services isolation points on pipe racks and underground at the demolition boundary;
  - Structural bracing and reinforcement locations.

— Register identifying radiation sources, hazardous materials, oil reservoirs, pressurized cylinders or anything else requiring removal in the area.

8.2.2. Shutdown, flushing and isolation

The process of shutting down plant can be staggered to enable flushing of the system and isolation of areas, prior to decontamination. In a uranium processing plant, water is typically pumped through the processing equipment, tanks and vessels and sent to a tailings management facility. Once flushed, the major assets requiring cleaning and decontamination can be shutdown, de-energized and isolated as required, to enable safe completion of decontamination.

8.2.3. Decontamination

In uranium mine and processing facilities radiologically contaminated assets will significantly influence the decontamination strategy. Radiation surveys are needed before decontamination to identify radiation contamination and after decontamination to clear plant and equipment. Decontamination may be as simple as flushing and cleaning systems with water to remove residual process material and clean surface contamination. Fixed radiation contamination requires additional decontamination measures, such as cleaning with chemical solutions (e.g. acids) or abrasive substances (e.g. sandblasting). For example, items that are exposed to process water are more likely to become contaminated and may require either high pressure washing or sand blasting. There can be a considerable cost in decontaminating plant or equipment below contamination levels that will allow items to be sold, removed off site or disposed of. A decontamination assessment of items that are radiologically contaminated can be completed to identify the decontamination method, and take into account the costs, versus the benefits.

The detailed process for decontamination varies dependent on local and national regulations and site specific conditions. Decontamination of assets in areas identified for demolition include the following activities:

— Decontamination of plant structures, equipment and auxiliary components;
— Preparation of equipment that will be disposed of whole and intact (e.g. for onsite burial);
— Identification of plant, equipment and components that can be sold;
— Identification of plant, equipment, components or materials that can be recycled, rather than disposed.

General guidance for the preparation and decontamination of equipment, in readiness for demolition, is outlined in Table 23.

TABLE 23. GUIDANCE FOR EQUIPMENT STATE IN PREPARATION FOR DEMOLITION

Equipment type	State in preparation for demolition
	Mechanical equipment
All equipment	Free of process fluids and solids. Where required to be removed from site, free of radioactive contamination or hazardous scale Free of hydrocarbons, hazardous dusts and gases. Free to the extent that surface films and the atmosphere in or around equipment is not ignitable Drained of water. Any residual water to be free of harmful concentrations of chemicals and radioactive material To be left in a free draining, depressurized and vented state Asbestos containing materials identified and removed or noted as still in place on hazardous materials register for demolition
Conveyors	Belts empty and tension released, with loose material cleaned away
Gearboxes, hydraulic systems, bearings	Drained of oil. A residual thin film of oil on internal surfaces may be acceptable
Piping and valves	For liquid and slurry handling systems, drain equipment or piping and then flush with water and drain
Hazardous substances vessels	To be left in a free draining (if above ground), depressurized and vented state. Any residual water to be free of harmful concentrations of chemicals and radioactive material
	Electrical equipment and instrumentation
All electrical equipment	Electrically de-energized and isolated
All cables	De-energized and air-gapped at feeder outside of demolition area and all cables air-gapped at a clearly visible location at the demolition battery limit Earth strap to be applied to any high voltage cables. All cables to be capped
Earthing/grounding system	Earthing/grounding system in demolition area to be separated from earthing/grounding system of energized sections
Transformers, motors, oil immersed equipment	Drained of oil. A residual thin film of oil on internal surfaces may be acceptable
Transformers	Drained of oil
All instrumentation	Electrically de-energized and isolated
Nucleonic instrumentation	Radioactive sources removed
Batteries	Removed from demolition area
	Radiation measurements
All assets	Radiation levels below the limits defined as per local jurisdiction requirements on the interior and exterior of all assets and in general demolition area
	Other
Septic tanks	Free of sewage. To be left in a vented state
Gas cylinders, fire suppressant, fire extinguishers	Depressurized
Air-conditioners, water coolers, ice makers	Free of refrigerant
Loose chemicals and reagents	Removed from demolition area
General housekeeping	Area tidy and general wastes separated for final disposal

8.2.4. De-energization and isolation (for demolition)

The de-energization and isolation of all plant and equipment within demolition areas need to be documented within the work packages and completed before demolition. At this stage, generally the only the equipment necessary to support the decontamination work is still operational, such as power, water, sump pumps, tailings pumps, etc. These assets need to be isolated to enable safe completion of demolition. Additional considerations for electrical and control systems, piping and structural plant are outlined below.

8.2.4.1. Electrical and control systems

Electrical and control systems need to be de-energized and isolated (in preparation for removal) in a way which allows power and controls systems to continue to service the auxiliary equipment that is required to remain operational during or after decommissioning. This may require the rerouting of high voltage and control cabling around the zone to be decommissioned. Other considerations for electrical and control systems include:

- — Maintenance of security systems (e.g. closed circuit television) in radiation controlled areas until all radioactive material has been cleared from the area or as per licence requirements (i.e. licences issued by nuclear security regulatory authorities).
- — Conducting an underground services survey within and around the boundary of the demolition area.
- — Electrical de-energization and isolation of all cables and electrical equipment in the demolition area by air-gapping of electrical feeders into the demolition area.
- — Electrical de-energization and isolation of all control valves, control cables and instruments by disconnecting cables from the energized source such that there is a visible gap. An alternative option would be to disconnect main feeder to the area.
- — Reviewing and updating interlock software and hardware, and any communication/control system interface points, to ensure de-energization and isolation of demolition area does not adversely affect continuity of services or works in areas outside of the demolition area.
- — Disconnection of all cables passing into the demolition area via cable trays or underground at a location clearly visible to the demolition contractor. The cables are de-energized by disconnection of feeders in substations; this work is intended to provide the demolition contractor assurance of the isolation.
- — Separating the earthing/grounding grids for energized and de-energized areas.

8.2.4.2. Piping

All piping passing into the demolition area is required to have positive isolation, preventing fluids or materials from entering into the demolition area. The positive isolation is to be easily visible and located as close as possible to the demolition area boundary and on the non-demolition side. The preferred method of positive isolation is a removable spool. If the upstream piping system is still energized (e.g. for services) then blanking plates need to be installed.

8.2.4.3. Structural stability

Design and installation of additional structural steel to stabilize structures that span across the demolition boundary may be required. Foundations need to be assessed to determine if excavation works will compromise their integrity.

8.2.5. Demolition

The handover of the demolition works requires all documentation, work packages and handover check sheets to be provided to the demolition team. Examples of demolition activities include:

— Visual demarcation of the demolition area boundary;
— Completion of the decommissioning work package and handover check sheet;
— Inspection of the demolition area to confirm completion of all activities in the decommissioning work package and documenting incomplete items;
— Approval of the completion of activities performed.

It may be necessary to expand the zones for radiation controlled areas during demolition such that movement of contaminated equipment and waste can be completed. Subsequently, this may then require removal and treatment of surface materials (e.g. soil) from these expanded controlled areas.

8.3. REHABILITATION

Mining activities may pose significant impacts and long term liabilities on the environment. Rehabilitation aims to minimize and mitigate all the environmental effects (radiological and non-radiological) of mining and processing related activities. A State's legislation and regulations govern the rehabilitation requirements associated with the closure and post-closure activities of a uranium mine and processing facility. Typically, the legislative framework requires environmental assessments during all phases of uranium mining and processing, from construction, through to operation and closure. Environmental assessments, environmental management programmes and closure plans inform the activities of rehabilitation.

Closure of a uranium mine or processing facility generally involves activities to reduce the radiation exposure or long term liability associated with radiological and chemical elements of concern. Examples of uranium production facilities requiring treatment are tailings management facilities, waste rock facilities, mine workings and water storage facilities.

At the time of rehabilitation, the decommissioning activities (e.g. decontamination and dismantling) are mostly completed, with the mine and processing facility infrastructure having been removed. As such, requirements for maintenance staff, and maintenance planning, scheduling and inspections will be significantly reduced.

Rehabilitation activities may take place over a number of years. Initially large heavy mobile plant is generally required for landform shaping and environmental restoration of the landscape. Long term rehabilitation activities typically require smaller fixed and mobile plant for environmental management and monitoring activities. As such, requirements for maintenance staff and maintenance planning, scheduling and inspections will be further reduced.

Maintenance of infrastructure generally require rehabilitation includes the following:

— Infrastructure:
  - Infrastructure including tanks, piping and equipment to support potable water, sewerage, storm water collection and water treatment facilities, to be maintained for the reduced size of the operation.
  - A change from using tanks to bunded fuel and oil pods which don't require fixed infrastructure and can be relocated with ease.
  - Maintenance workshops will still be required to service pumps, land management tools, service vehicles and heavy equipment.
  - Camp and mess facilities.
  - Nursery and watering system for revegetating and rehabilitation.

    - Water collection and sump management, particularly within redundant or decommissioned areas where sump systems may be disconnected or out of service. This may require portable pumps to be set up, monitored and started when required to transfer water.
    - Road maintenance.
    - Reduced capacity office areas and warehouses.
- Electrical and control system:
    - High voltage and low voltage systems;
    - Lighting;
    - Control system and monitoring systems, particularly for alarms.
- Power generation:
    - Power stations or generators, either with fixed or portable fuel supply systems;
    - Transition from electrical or fuel based systems to solar system, to power smaller plant.
- Vehicles:
    - Light vehicles for transportation;
    - Heavy mobile equipment for land-form design;
    - Land management equipment, e.g. tractors, ride-on mowers, all-terrain vehicles.

The location of a uranium mine and processing facility is constrained by the location of the uranium deposit. For this reason, site supporting facilities (e.g. roadways, transportation links, electrical grids, water supply infrastructure) are often developed to support the operation. A uranium mine or processing facility at the end of its operational life, will generally require the same site supporting infrastructure for closure activities. For example, electrical grids to provide power for water treatment facilities and camp facilities for staff completing closure activities. Typically, site supporting infrastructure will not remain after closure, as assets are decommissioned and demolished. However this changes if the government or community identifies key assets that they wish to maintain or utilize (e.g. power generating assets, communication towers, water distribution pumps/pipelines) [24], requiring the owner/operator to consider the handover of ownership of site supporting assets. From an asset management perspective, the owner/operator is expected to ensure that assets being retained for handover are maintained in a condition that does not pose an unreasonable risk to people or the environment.

8.4. RETURN TO INSTITUTIONAL CONTROL

Once all required closure activities have been completed in accordance with the state, provincial and/or national regulations, the owner/operator will be interested in divulging ownership and responsibility of the site that contained the uranium mine and processing facility. This process in many jurisdictions is referred to as returning the site to an institutional control programme, for post-closure management of the decommissioned mine and/or processing facility. For a uranium mine and processing facility owner/operator to enter into the institutional control programme they need to ensure the following:

- The site does not pose an unreasonable risk to the environment;
- The site does not pose and unreasonable risk to the safety of the public;
- The site does not pose a risk to national security.

Further, the institutional control programme defines conditions to be implemented (i.e. criteria) and the process for long term monitoring and maintenance of the decommissioned uranium mine and processing facility. Following regulatory assessment of the requirements for long term management of the site, the owner/operator is required to provide the province or state with sufficient funds to conduct long term monitoring and associated maintenance.

Once the conditions of the institutional control programme (both site based activities and financial commitments) are met, the owner/operator then may be released from any regulatory oversight of the site as well as any provincial, state and national financial obligations such as taxes and royalties. In essence, the total custodial responsibility of the site is released back to the province or state. At this point, the province or the state will be responsible for the long term oversight and maintenance of the property and any remaining assets.

From an asset management strategy perspective, the owner/operator can now consider this to be the end of the life cycle of the uranium mine or processing facility and cease any asset management strategies.

APPENDIX I.

BOW TIE ANALYSIS EXAMPLES FOR AMMONIA, SULPHURIC ACID AND SOLVENT EXTRACTION

Examples of process safety bow tie analysis for ammonia, sulphuric acid and solvent extraction are shown in Tables 24–26, respectively.

TABLE 24. EXAMPLE BOW TIE ANALYSIS FOR AMMONIA FOR A URANIUM PROCESSING FACILITY

Hazard: Ammonia								
Threat	**Preventative barrier 1**	**Preventative barrier 2**	**Preventative barrier 3**	Top Event: Loss of Containment of Ammonia	**Recovery barrier 1**	**Recovery barrier 2**	**Recovery barrier 3**	**Recovery barrier 4**
Ammonia release during unloading activity or overfilling of storage vessel	**High level alarm:** Accurate tank level and alarm sounds or high level alarm	**Unloading authorization:** Unloading will only commence if operator confirms driver checks completed	**Unloading operator competency:** Delivery driver trained and competent		**Ammonia open path detector:** Ammonia detectors will detect ammonia release and sound alarm	**Ammonia leak Detector:** Ammonia detectors will detect ammonia release and sound alarm	**Ammonia alarm siren:** Regular testing of siren	**Emergency response:** Trained and competent emergency response with right equipment
Corrosion loss of Integrity	**Vessel mechanical integrity:** Vessels comply with AS/NZS 3788 and remain fit for service	**Pipeline mechanical integrity:** Pipework complies with AS/NZS 3788 and remains fit for service	**Operator checks:** Visual inspection of storage vessels and pipework will identify early signs of failure					
Overpressure of storage vessel or piping system	**Pressure relief valves:** PRV's will activate pressure to prevent overpressure	**Fire break:** Adequate fire break in place around storage vessels and piping						
Ammonia release during operational activity	**Area operator competency:** Operators are trained and competent to work in the ammonia plant							

TABLE 25. EXAMPLE BOW TIE ANALYSIS FOR SULPHURIC ACID FOR A URANIUM PROCESSING FACILITY

Hazard: Sulphuric Acid							
Threat	**Preventative barrier 1**	**Preventative barrier 2**	**Preventative barrier 3**	**Preventative barrier 4**	Top Event: Loss of Containment of Sulphuric Acid	**Recovery barrier 1**	**Recovery barrier 2**
Corrosion of a storage tank	**Acid quality:** Acid quality meets specifications	**Tank mechanical integrity:** Tanks comply with API STD 653 and remain fit for service	**Tank mechanical integrity:** Visual inspections to identify obvious damage or corrosion			**Distribution pump emergency stop:** Remote emergency stop for distribution pumps	**Secondary containment:** Bund capacity remains at 110% of largest tank
Corrosion/erosion of distribution pipework	**Pipeline mechanical integrity:** Pipeline complies with AS/NZS 3788 and remains fit for service						
Sulphuric acid spill during unloading	**Engineering design of unloading hoses:** All hoses are manufactured to specifications	**Road tanker mechanical integrity:** Tankers maintained to local codes	**Unloading hose mechanical integrity:** Hoses, materials and fittings meet specifications. Hoses are in good condition	**Training and competency:** All delivery drivers are trained and competent to unload			
Overfilling of a storage tank	**Automated unloading pump shut off:** Tank level will not exceed safe fill level, unloading pumps automatically shut down	**Overfill protection:** Secondary, independent level sensor will override control system and shut down unloading pumps	**Unloading authorization:** Unloading will not commence if the tank level is too high	**Training and competency:** All delivery drivers are trained and competent to unload			

TABLE 26. EXAMPLE BOW TIE ANALYSIS FOR SOLVENT EXTRACTION FOR A URANIUM PROCESSING FACILITY

Hazard: Solvent Extraction						
Threat	**Preventative barrier 1**	**Preventative barrier 2**	**Preventative barrier 3**	Top Event: Fire in Solvent Extraction	**Recovery barrier 1**	**Recovery barrier 2**
Ignition from lightning strike	**Lightning earthing system:** Lightning protection system will transfer energy to earth				**Fire protection system:** Fire protection system activates to detect and/or extinguish any fire and remove fuel from the fire zone	**Operator training and competency:** Operators are trained and competent to respond to a fire
Bush fire	**Fire break:** Separation between a fire and the SX Plant will reduce the risk of SX Fire	**Operations preparation:** Plant and response equipment is prepared for use if required				
Ignition from equipment fault	**Equipment electrical integrity:** Electrical circuit will trip if overload or short circuit is detected	**Equipment mechanical integrity:** Equipment does not become faulty and lead to and ignition source	**Equipment mechanical integrity:** Visual inspections, leaks and equipment faults are rectified promptly			
Ignition from static	**Earthing of portable pumping and pipework:** Static electricity cannot accumulate and provide an ignition source					

APPENDIX II.

MANAGEMENT OF CHANGE PROMPT LIST

An example list of prompts for MOC are provided in Table 27, which may be considered for the development of an MOC form for a uranium mine and processing facility.

TABLE 27. MANAGEMENT OF CHANGE PROMPTS

MOC prompt	Example
Health, safety, environment and communities	
Could the change have a potentially significant health, safety, environment or community risk?	
Could the change impact health?	Consider biological vectors, vapour, gas, noise, light, heat, dust, hazardous substances, ergonomics, manual handling, vibration, thermal energy, legionella, radiation, etc.
Could the change impact the environment?	Consider liquid, solid or gaseous waste, energy use, greenhouse gas, potential for leaks, hazardous substances, land disturbance, air quality, noise, vibration, weeds, vermin, etc.
Could the change impact safety?	Consider isolations, electrical safety, vehicles and driving, working at heights, confined spaces, cranes and lifting, aviation, etc.
Could the change impact the indigenous communities and/or region?	Consider stakeholder relationships, cultural heritage sites, air quality, water quality, traffic, power supply, etc.
Does the change need to be communicated to stakeholders?	Consider national parks, regional council, local utility provider, school, government body, etc.
Does the change comply with the operation's authorization or licence to operate?	Consider legislation, regulations and stakeholder requirements
Could the change have the potential to impact the reputation of the operation or shareholders?	Consider contingency planning, business resilience, emergency response, media attention, etc.
Does the change involve the introduction of new hazardous substances or changes to the way existing hazardous substances are used?	
Could the change require new documents or amendments to, or removal of existing documents?	Management plans, monitoring programmes, permits, safe operating procedures, forms, templates, plans, etc.
Is the change associated with hire/purchase of new plant or equipment?	Consider if safety devices and site requirements are met, emergency stops installed, electrical earthing, vehicle rollover protection, communication equipment, registered equipment, etc.
Radiation protection	
Could the change affect radiation protection programme including dose limitation ?	Work involving plant dust extraction/ventilation systems, work that may change controlled area boundaries/demarcation, work where radon gas could be present, work near the neutral thickener or return water lines or work that may generate dust, etc.
Could the change involve plant or equipment near sources of uranium ore concentrate (product)?	Work on any plant or equipment located in the precipitation, calciner or product packing buildings
Could the change release radioactive material to the environment?	Work that could release radon, dust, process water, uranium ore concentrate (product)
Could the change affect the transport of product from the mine?	Work on the drum conveyor, work in the product warehouse, work in the product yard, work involving product containers, work that affects the transport of product, etc.

TABLE 27. MANAGEMENT OF CHANGE PROMPTS (cont.)

MOC prompt	Example
Process safety	
Does the change require the introduction of a 'temporary building' at the operation?	Temporary buildings include trailers, container boxes, semi-trailer trucks, portable building modules and tents that are commonly used for accommodation for maintenance shutdowns, temporary office, training facility and change rooms, etc.
Could this change affect process safety?	Work involved with high pressure vessels, reactive or hazardous chemicals and its storages, bulk process material storages, potable water system and tailings facilities, etc.
Does the change require a pre-start safety review after its completion?	Verification of construction, equipment and other components, employee training, handover of responsibilities, etc.
Does the change impact the process safety hazard packages?	Process safety risk assessments, bow ties, control activity sheets, verification activities, etc.
Emergency and security	
Could the change affect fire detection, fire prevention or fire protection?	Work that may disable or isolate alarms systems, fire suppression systems. Works that may trigger false alarms. Works occurring in new or remote areas that do not have fire systems in place.
Could the change affect emergency response?	Consider escape routes, rescue equipment, communication equipment, etc.
Could the change affect compliance to the operation's security obligations?	Consider access to site, ammonia plant, SX plant, precipitation, calciner or product packing areas, etc.
Closure	
Does the change impact closure planning or schedule?	Consider process water inventory management, tailings quantity and quality, waste management, etc.
Does the change require a new evaluation or change to an existing best practicable technology assessment?	Consider regulatory approvals for environmental assessments, closure plan or licence conditions for closure. Best practicable technology is required to support a regulatory application involving a change to the process or waste management system.
Human resources	
Could the change impact on human resources?	Consider role descriptions, role accountabilities, human resources systems, personnel, organization structures, etc.
Could the change impact training?	Consider inductions, training materials, learning programmes, etc.
Information technology	
Could the change impact communications or networks?	Consider technical systems, support software and tools, wireless networks, copper and fibre-optics etc.
Does this change impact cyber security and remote access?	Consider information management and security guidelines, electronic documents, databases, digital control systems, programmable logic controller, etc.
Does the change include removal of applications and/or approved software?	Contents of software may need export in approved format, classified in accordance with retention schedule and decommission of application from server.
Does the change impact data privacy or file sharing?	Consider information management and security guidelines, permissions, etc.
Processing operations	
Could the change impact control systems and instrumentation for plant monitoring or control?	Consider digital control systems, programmable logic controller, plant layout, equipment positioning, operations during routine, abnormal and emergency situations, etc.
Could the change affect upstream or downstream processes?	Consider flow rates, composition, alarms, isolations, etc.

TABLE 27. MANAGEMENT OF CHANGE PROMPTS (cont.)

MOC prompt	Example
Processing operations	
Could the change affect plant startup/shut down activities?	
Could the change affect process composition?	Consider, reagents, reactions, temperature, pH, etc.
Mining	
Does the change cause a deviation from standards for geotechnical stability?	
Does the change impact reserve estimates?	
Does the change impact traffic flow or management?	
Does the change require new or updated operating procedures?	
Could the change exclusively impact pit permit holders?	Access to active mining areas, restricted areas, requirements for fencing or other barriers, etc.
Maintenance	
Does the change require preventive maintenance to be scheduled?	Including cleaning, inspections, maintenance procedures, etc.
Does the change require new or updated maintenance requirements?	Consider access, lighting, isolations, ergonomics, etc.
Could the change affect procurements? Do spares need to be criticality assessed?	Consider spares required and stocks, bill of materials required to be updated, warranty conditions, etc.
Water systems	
Does the proposal include connection to or work on the site potable water system?	Consider all water infrastructure (pipelines, pumps, non-return valves, monitoring equipment) etc.
Does the change increase or decrease site water consumption?	Consider extraction and discharge licence requirements, water balance models, water availability, etc.
Does the change result in water of a different class (raw, process, potable, treated) being used in an area?	Consider all water class interactions.
Does the proposed change have any impact on the water quality which is discharged from an area?	Consider operation parameter, heath regulations, national water quality guidelines, licence conditions, etc.
Does the proposed change modify the discharge locations of any effluent or drainage systems?	
Engineering design	
Could the change impact electrical and instrumentation engineering?	Consider isolations, alarms, static electricity, lighting protection, ingress protection ratings, protective devices, control system modifications, wiring, electrical equipment, etc.
Could the change impact reliability and mechanical engineering?	For example, material changes, equipment, flow rate, component specification, part size, wear components, stored energy, isolations, registered plant, loading, structural integrity, corrosion, or other analogous changes
Quality assurance documentation	
Does the change require a handover agreement?	Consider handover between departments, between operator and contractor, and operator and other external party (e.g. institutional control programme in closure), etc.
Does the change require use of design standards and specifications?	Consider regulatory requirements and national guidance, international standards and guidance of the operating organization, etc.
Does the change require use of supplier or manufacturer information?	Consider material data reports, data sheets, design specifications, performance curves, calculations and functional descriptions, etc.

TABLE 27. MANAGEMENT OF CHANGE PROMPTS (cont.)

MOC prompt	Example
Quality assurance documentation	
Does the change require use of instruction manuals, procedures, or plans?	Consider installation instructions, inspection test plans, repair procedures, inspection test plans, etc.
Does the change require a factory acceptance test?	
Could the change impact statutory equipment registration e.g. pressure vessels, lifting equipment?	Consider details of equipment design, equipment registration with local authority, inspection and maintenance logs, etc.
Does the change require retention of records for operational, statutory, auditing or other analogous purposes?	Consider commissioning records, test certificates, tests sheet, test records, training record, qualifications certificates, material numbers, consumables tracking records, etc.
Drawings	
Does the change require electrical and instrumentation drawings to be updated?	Single line diagram, schematics, wiring diagrams, etc.
Does the change require mechanical, civil, structural, piping drawings to be updated?	General arrangement and details, process flow diagrams, isometric drawings, etc.
Does the change require piping and instrumentation diagrams to be updated?	
Contingencies	
What contingencies are in place in the event the change cannot be fully implemented?	If the change cannot be completed as planned, consider all possible risk and impacts that might occur. Describe contingency plans in place to control any impacts or risks identified

Note: Other prompts will be necessary dependant on the organization.

APPENDIX III.

LESSONS LEARNED FROM URANIUM PROCESSING EQUIPMENT FAILURES

A series of real-world case studies of equipment failures in uranium processing facilities are provided by industry experts, summarizing lessons learned through investigation of the failures. The examples provided are not exhaustive but demonstrate the range of factors that can affect equipment failure, including planning, maintenance, operations, organization, change management and data management. Basic root causes (or findings) which led to the failures of the equipment are described. The corresponding corrective actions that were implemented to prevent reoccurrence of similar failures are included. In most cases, the equipment failures could have been avoided if a rigorous maintenance strategy had been implemented.

Note: This appendix contains technical language and references to specific industrial equipment that is not explained in detail.

III.1. LEACH TANK FAILURE

III.1.1. Consequence

Uncontrolled release of process solution and significant loss of production.

III.1.2. Background

Leakage was detected on a leach tank shell. While in the process of the tank being bypassed, the tank sheared open resulting in a catastrophic failure.

III.1.3. Findings

- Management of change:
  - Modified/retrofitted vertical stabilizing rails on external of tank were not recorded and therefore the significance of the design change and maintaining its integrity had not been assessed or understood.
- Oversight on signs of failure:
  - No focus on preventing external tank corrosion. The tank's external surface was not corrosion protected.
  - Deep vertical external corrosion channels that appeared on the tank were not risk assessed to understand the impact on the tank structural integrity.
- Maintenance procedures:
  - Inadequate repair procedures and quality. Issues related to methods for thickness testing, lack of spark testing rubber lining, no inspection and leak checks of tank floors and poor quality assurance of repair methods.

III.1.4. Corrective actions

- Review maintenance strategy to ensure focus on preventive maintenance;
- Perform trend analysis to determine priority areas;
- Implement MOC procedure for all process and structural changes;
- Establish detailed maintenance schedules, procedures and quality checks for tank internal and external inspections, testing and repairs;
- Mandate the capturing of maintenance history in the CMMS.

III.2. CONVEYOR PULLEY FAILURE

III.2.1. Consequence

Unscheduled and rushed maintenance work due to bearing failure.

III.2.2. Background

A conveyor pulley was found to be noisy and badly vibrating just prior to a planned plant shutdown. Vibration analysis was conducted. An outer raceway defect was detected, resulting in the pulley being changed out.

III.2.3. Findings

- Inspections and a mechanical running inspection ten days prior to the shut did not detect the defect. Given the size of the defect it ought to have been picked up allowing more time to plan the changeout.
- The inner ring had spun on the adaptor sleeves. There was fretting corrosion between the shaft and sleeve leading to a loss of clamping force. The cause of the fretting was not able to be determined. Key measurements were not taken due to the rushed repair of this pulley. The most likely cause would be undersized shaft or lock nut not tight enough.
- The pulley was not overhauled to the required standards. There was only one spare component available for 22 locations which resulted in this overhaul being rushed and short cuts taken.
- Primary maintenance tasks took too long to schedule. Three primary maintenance tasks that had been planned took 62 days to schedule and execute the work.

III.2.4. Corrective actions

- Ensure production operators and inspectors investigate and report strange sounds;
- Carry out a detailed inspection programme to ensure correct quality of spares and pulleys;
- Create and track a key performance indicator for primary maintenance scheduling and execution so that the length of time from failure identification to the time the item is repaired can be tracked and actively improved upon.

III.3. CLARIFIER VALVE FAILURE

III.3.1. Consequence

Uncontrolled release of process liquor into bermed area and loss of production.

III.3.2. Background

The failed valve was observed to have split the valve seat which then became the leak path for process liquor to make contact with the cast iron valve body which was susceptible to corrosion. Consequently, the valve body was corroded through, resulting in an uncontrolled release of process liquor into the catch basin.

III.3.3. Findings

- The valve installed was a butterfly valve with Buna-N (nitrile) rubber that was incompatible with the aqueous solution;
- The valve body was cast iron and susceptible to corrosion once the seat failed.

III.3.4. Corrective actions

- Replace the failed valve and the adjacent one with a Teflon lined ethylene propylene diaphragm valve.
- Implement quality control steps at installation. Include a step in the work order to check equipment specification.

III.4. SAND FILTER DISPLACEMENT DRAIN VALVE FAILURE

III.4.1. Consequence

Uncontrolled release of process liquor into bermed area and loss of production.

III.4.2. Background

Failed valves were observed to have split valve seats which then became the leak path for process liquor to make contact with the cast iron valve body, which was susceptible to corrosion. Consequently, the valve body was corroded through, resulting in an uncontrolled release of process liquor into the bermed area.

III.4.3. Findings

- Sand particles (from the backwash process) are sometimes present in the displacement drain line.
- Sand builds up behind the valve disc/shaft and grinds away the valve seat when the valve is opened.
- Installation instructions required that when the material handled is a slurry and solids are likely to fall out of suspension, the valve is installed with the valve shaft horizontal, so that the disc will open in the downstream direction at the bottom.
- Valves were not installed according to installation instruction.

III.4.4. Corrective actions

- Tilt the valve such that the shaft is in a horizontal plane to prevent sand abrasion between the disc;
- Ensure valve opens downstream so particle build up behind the disc will flush out when opened;
- Familiarise workforce with operating specifications and utilize these.

III.5. CLARIFIER BOGGED RAKE

III.5.1. Consequence

Loss of production.

III.5.2. Background

The clarifier rake stopped and remained in a stationary position for approximately four hours. When the plant operators became aware that the rake was stationary, a restart was attempted. By this time, the rake was bogged and the unit tripped on over-torque.

III.5.3. Findings

- The torque reading on the DCS was not accurate and the operations staff were not watching it, relying on manual checks. There was also no alarm for loss of torque.
- A review of the trend data for the rake torque indicated that the rake stopped in the afternoon (as indicated by the small dip in torque). The cause of this stoppage was unknown, but the brief failure was enough to bog the rake and it was unable to start again. No alarm was triggered or logged that would have made the operator aware that the unit had stopped.

III.5.4. Corrective actions

- Replace torque sensor on rake;
- Implement DCS alarm for rake stoppage;
- Inspect DCS configuration for similar issues on other drives and areas.

III.6. CLARIFIER PLANETARY DRIVE FAILURE

III.6.1. Consequence

Loss of production.

III.6.2. Background

Following from the failure described in Section III.5, failure of the planetary drive output shaft lower bearing was due to excessive radial load on the bearing, induced by a bending load on the rake assembly.

The rake assembly (shaft and arms) is completely supported by the planetary drive. Due to the physical configuration of the bearings only small bending loads on the shaft can be withstood.

Under normal operating conditions the bearings are subjected only to vertical loads. However, if the rake is bogged or partially bogged when the drive is restarted, there is potential for bending loads to be applied to the bearings. This increases the radial load on the lower bearing significantly, and as was observed in this failure, distortion of the rake structure and subsequent bearing failure.

III.6.3. Findings

- A process upset investigation was completed for a rake bogging event two days prior to this failure.
- The clarifier was restarted while the tank rake was still partially bogged.

III.6.4. Corrective actions

- Review restart methods for bogged thickener rake, ensuring the rake is lifted to full height before starting.

This failure demonstrates how a failure of a piece of equipment can lead to further issues and subsequent failures as a result.

III.7. ACID PIPING VALVE FAILURE

III.7.1. Consequence

Uncontrolled release of acid.

III.7.2. Background

Premature failure of acid control valves in the leach area. The valves were leaking acid from the gland seal of the plug shaft.

Valves are constructed predominantly of 316 stainless steel and utilize a polytetrafluoroethylene based composition (Teflon) for the seats and seals, both of which are an appropriate material for 98% sulphuric acid. The valve sealing arrangement has upper and lower sets separated by a spacer and tensioned by a gland flange. Polytetrafluoroethylene is subject to creep, and it is specified by the OEM that the gland flange is to be checked, and tightened if necessary, each month at a minimum.

III.7.3. Findings

- Inadequate commissioning of valves.
- Inadequate inspection instructions and maintenance data for valves. Even though the valves are tested and certified from the factory, there is still a need to adjust the gland packing once installed as the polytetrafluoroethylene will creep over time and lose its sealing capability. Maintenance personnel were not aware of the requirement to adjust the gland packing post-installation. Commissioning instructions and standard operating procedures outlining this requirement are not available to maintenance personnel.
- Regular checks and adjustments of the valve seals are not being completed as specified by the manufacturer. If the valve is noted to be leaking, the valve is washed with potable water as soon as possible and the gland packing tightened to stop the leak. There is a total of 15mm of adjustment available in the valve seal, and only once the maximum adjustment has been carried out and a leak remains will the valves be replaced.

III.7.4. Corrective actions

- Develop maintenance procedures and commissioning documentation;

— Daily checks to be conducted by production operators for acid leaks and recorded on log sheets;
— Fortnightly maintenance activities to check, and tighten when required, gland packing;
— Safety instructions for this activity to be treated as aggressive release;
— OEM to provide information for training session to maintenance personnel.

III.8. ACID PIPING: PUMP MECHANICAL SEAL FAILURE

III.8.1. Consequence

Uncontrolled release of acid.

III.8.2. Background

An aged acid pump was replaced. However, after few days of function testing the mechanical seal of the pump failed and acid leaks were reported.

III.8.3. Findings

— The seal used in the failed valve had 316 stainless steel for the seal body but had carbon steel faces. The OEM specification for this seal is not compatible with 98% sulphuric acid conditions.
— The supplier of the original pump had modified the pump from the manufacturer for it to be suitable for acid service (with the correct mechanical seal) prior to supplying the pump for installation. They did not however replace or update the specification plate on the pump itself.
— When the new pump was ordered, only the specification on the pump nameplate was referenced. The fluid type or application was not discussed.
— The original pump without the modified mechanical seal was therefore installed.

III.8.4. Corrective actions

— Planners to consult with engineers to review specifications prior to ordering equipment.
— Equipment specifications to be updated in the 'bill of materials' so equipment ordering can be accurate.

III.9. AMMONIA LEAK DUE TO GASKET FAILURE

III.9.1. Consequence

Uncontrolled release of ammonia.

III.9.2. Background

While charging the ammonia system, ammonia was detected leaking from the ammonia system in the precipitation area.

The suspected leaking valves were tightened, leading to further leaks.

III.9.3. Findings

- During a valve change out, Teflon gaskets were installed instead of the spiral wound gaskets specified.
- No quality control checks were in place for the changed out valves or gaskets installed.

III.9.4. Corrective actions

- Implement quality control documentation for gasket installation outlining the type of gasket to be used and installation procedures including alignment, torque settings, bolt specifications and catalogued material numbers.

III.10. LEACH TANK AGITATOR HUB AND STUB FAILURE

III.10.1. Consequence

Loss of production and high risk of damage to tank.

III.10.2. Background

- The leach tank agitators have three blades at the base and partway up the agitator for a total of six blades (in offset positions).
- Abnormal heavy agitation of the tank led to the tank level being dropped for further investigation where a blade was found to be partially 'torn' from the shaft and twisted causing the flat area to be perpendicular to the direction of flow creating the increased agitation in the tank.
- The blade soon after completely tore off and fell to the bottom of the tank. The tank was drained, and the blade recovered for analysis.

III.10.3. Findings

- A fatigue fracture initiated at the internal diameter of the hub at a region which exhibited manufacturing related surface flaw acted as stress concentrators. The surface flaw, which measured to be approximately 100 mm wide was likely the result of excessive vibration and chattering during machining.

III.10.4. Corrective actions

- During fabrication, all agitator components require visually inspecting following machining to determine the presence of defects which could act as stress concentrators.

III.11. LEACH TANK FAILED AGITATOR SHAFT

III.11.1. Consequence

Loss of production.

III.11.2. Background

The agitator shaft within the leach tank was in operation for approximately 12 months before the shaft failed at the top stub and fell into the leach tank. The shaft was powered by a 55 kW drive through a reduction to achieve 13.8 revolutions per minute. The shaft and flange were manufactured from 2205 duplex stainless steel.

III.11.3. Findings

The investigation included visual examination, chemical analysis, material identification, tensile and hardness testing, scanning electron microscopy, microstructural evaluation and torque calculations.

- The heat effected zone and weld structure between the shaft and top flange exhibited defects including both lack of side wall fusion and welding flux or slag inclusions measuring 45 mm in length and 1.5 mm wide.
- These defects were considered likely to have acted as geometric stress concentrations in the initiation of the fatigue fracture.
- The defects found were not consistent with the quality level required for applications subject to cyclic loading.

The nature of the failure suggested that the agitator shaft ought to have been manufactured to fatigue purpose welding codes such as Ref. [89, 90]. The use of NDT is expected to ensure that the incidence of internal weld defects is reduced and the likelihood of further weld failures is minimized as a result.

III.11.4. Corrective actions

- As the failure had originated from internal weld defects it was identified that the future agitator shaft fabrication welds be subjected to NDT for internal defects as part of the manufacturing process.
- Specific requirements are to be reviewed before requesting the manufacture of new shafts. These requirements could include but not be limited to the following:
  - Request as-built drawings which detail joint configuration, welding process and applicable welding standards;
  - Request that future shafts are subject to additional NDT, such as X ray radiography or Time of flight diffraction, and are assessed and complied with either Ref. [90] or Ref. [89].

APPENDIX IV.

TACTICS FOR ROTATING EQUIPMENT

The FMEA process described in Section 4.4 lends itself well to the development of maintenance tactics for rotating equipment. Tables 28–30 provide examples of degradation mechanisms/failure modes for various types of rotating equipment, including pumps, conveyor belts, gearboxes and couplings. The tables also include associated tactics to prevent the failure from occurring. In some of the examples the warning effects that may be observed as the asset wears and approaches failure have been included. Finally, the tables summarize the associated maintenance task description based on the failure and a brief overview of the repair required.

TABLE 28. TACTICS FOR COUPLINGS

Function	Functional Failure	Failure Mode	Failure Cause	Effects of Failure	Failure Pattern	RPN Number (S×O×D)	Maintenance Actions	Frequency
To transmit power	Failure to transmit power	Wear and tear	Continuous operation	Loss of efficiency, increased vibration	Wear-out	420 (7×7×8)	Inspect and replace coupling	Every 6 months
		Misalignment	Improper installation	Premature wear, noise, equipment damage	Random	560 (8×7×10)	Realign shafts, inspect coupling	As needed
		Elastomeric element failure	Ageing or chemical exposure	Loss of torque transmission, possible disconnection	Early	450 (9×5×10)	Inspect and replace elastomeric elements	Annually
		Coupling bolt failure	Over-torque, fatigue	Coupling failure, potential for shaft separation	Random	630 (9×10×7)	Regular inspection, torque to specifications	Every 6 months
		Corrosion	Environmental exposure	Reduced strength, failure to transmit power	Increasing	320 (8×8×5)	Regular inspection, corrosion protection	Annually
		Lubrication failure	Inadequate maintenance	Increased wear and tear, overheating	Random	400 (8×5×10)	Regular lubrication, maintenance	Quarterly
	Loss of alignment	Shaft misalignment	Improper installation	Increased wear, vibration, and potential failure	Random	490 (7×7×10)	Regular alignment checks	Every 6 months
		Thermal expansion	Temperature variations	Misalignment, uneven wear	Increasing	270 (6×9×5)	Design consideration, regular checks	Annually

TABLE 29. TACTICS FOR CONVEYOR BELTS

Action	Functional Failure	Failure Mode	Failure Cause	Effects of Failure	Failure Pattern	RPN Number (S×O×D)	Maintenance Actions	Frequency
To transport material	Inability to transport material	Belt wear or tear	Continuous use	Reduced efficiency, potential belt snap	Wear-out	560 (8×10×7)	Inspect and replace belt	Every 6 months
			Abrasive materials	Accelerated wear, inefficiency	Random	420 (7×10×6)	Regular cleaning, check for sharp objects	Quarterly
		Misalignment	Improper installation	Uneven wear, belt tracking issues	Random	630 (9×10×7)	Realign conveyor belt	As needed
			Structural shifts	Belt drift, increased wear	Increasing	450 (9×5×10)	Inspect structure, realignment	Annually
		Motor failure	Electrical fault	Conveyor stops, production delay	Random	360 (6×6×10)	Electrical inspection, replace motor	As needed
			Overheating	Reduced motor life, stoppage	Random	288 (6×8×6)	Install cooling systems, regular maintenance	Every 3 months
		Roller/bearing failure	Wear and tear	Increased friction, noise, vibration	Wear-out	504 (8×9×7)	Inspect and replace rollers/bearings	Every 6 months
			Lack of lubrication	Premature wear, increased power consumption	Early	320 (8×8×5)	Lubricate regularly	Quarterly
	Loss of synchronization	Control system failure	Software glitch	Miscommunication, operational errors	Random	200 (5×8×5)	Regular software updates, checks	Every 6 months
			Sensor malfunction	Incorrect speed, stoppage, material spillage	Random	480 (8×10×6)	Inspect and replace sensors	Quarterly
		Blockages	Accumulated debris	Material spillage, belt stoppage	Increasing	600 (10×10×6)	Regular inspection and cleaning	Monthly
		Emergency stop activation	Unexpected object	Immediate halt, potential material damage	Random	250 (5×10×5)	Regular monitoring, employee training	As needed

TABLE 30. TACTICS FOR GEARBOXES

Function	Functional Failure	Failure Mode	Failure Cause	Effects of Failure	Failure Pattern	RPN Number (S×O×D)	Maintenance Actions	Frequency
To transmit power	Ineffective power transmission	Gear wear	Continuous operation	Increased noise, decreased efficiency	Wear-out	525 (7×7×9)	Inspect and replace gears	Every 6 months
			Lack of lubrication	Premature wear, overheating	Early	480 (8×8×7)	Regular lubrication	Quarterly
		Gear tooth breakage	Overload	Sudden transmission failure, potential secondary damage	Random	640 (8×8×10)	Inspect gears for damage, ensure proper load	As needed
			Material defect	Unexpected failure	Early	360 (6×6×10)	Quality control, replace defective parts	As identified
		Bearing failure	Wear and tear	Increased vibration, noise, heat	Wear-out	504 (8×9×7)	Inspect and replace bearings	Every 6 months
			Inadequate lubrication	Overheating, premature failure	Random	450 (9×5×10)	Regular lubrication, monitor lubricant quality	Quarterly
		Seal leakage	Ageing or wear	Oil leakage, contamination	Wear-out	315 (7×5×9)	Inspect and replace seals	Annually
			Improper installation	Premature failure, leakage	Random	280 (7×8×5)	Ensure proper installation, training	As needed
		Misalignment	Improper installation	Increased wear, potential for gear/bearing damage	Increasing	400 (8×5×10)	Realign gearbox, regular checks	Annually
			External forces	Vibration, alignment issues	Increasing	225 (5×9×5)	Secure mounting, vibration dampers	As needed
	Loss of lubrication	Oil contamination	External debris	Increased wear, decreased life expectancy	Random	540 (9×6×10)	Filter and replace oil, seal checks	Every 3 months
		Oil degradation	High temperature	Reduced lubrication quality, increased wear	Increasing	480 (8×8×7)	Monitor temperature, replace oil	Semi-annually

APPENDIX V.

RISK BASED ASSESSMENT PROCESS

V.1. OVERVIEW OF RISK BASED ASSESSMENT

The RBI methodology was developed by the oil and gas industry as a process to gain a deeper understanding of the damage mechanisms and controls required to manage tanks, pipes and pressure vessels. The method is considered to be more in depth, when compared to the general requirements that are outlined in industry standards or specifications and can therefore be used to justify an extension to those maintenance requirements and intervals. The deliverables of the RBI assessment are:

— A transparent and auditable report: The methodology is documented and application of the described methodology is traceable in the individual equipment reports.
— Targeted inspection and test plans: Equipment inspection plans are developed with tests that are focused on the damage mechanisms identified in the RBI assessment.
— A basis for discussion with statutory authorities: Extension of inspection frequency may require consultation with regulatory bodies. The governing statutory requirements may allow for an extension of inspection intervals when an RBI assessment is conducted, and the extension is justified.

The methodology described in this appendix is intended to optimize the availability of assets and ensure that the integrity and safety of equipment is addressed. It describes the methods used to determine the 'likelihood of failure' and the 'consequence of failure' of equipment, and hence risk.

RBI methodologies can range from qualitative, with a reliance on engineering judgment, to quantitative, where all aspects of likelihood and consequence are mathematically modelled. The approach used herein is semi-quantitative. Mathematical modelling is used, where practicable and combined with engineering judgment. This semi quantitative process is usually only performed on assets that have high maintenance costs or those which are high risk due to the time and cost in completing the RBI process.

This RBI methodology can be applied to tanks, pressure vessels, piping and associated safety controls (e.g. pressure relief devices). The methodology considers internal and external damage mechanisms that could affect the integrity of the equipment and lead to a loss of containment. It does not consider rotating equipment, internal fittings, electrical systems or infrastructure attached to the asset. Maintenance tactics for this equipment are developed following the process outlined in Section 4.3.

Risk prioritization enables the optimum level of inspection, and inspection effectiveness, to maintain the risk of failure at its present level or to reduce the risk if the present level is unacceptable.

Equipment integrity is maintained by executing inspections and monitoring processes at the defined levels. This includes identifying and utilizing appropriate inspection techniques with the appropriate coverage and appropriate frequencies. Inspection techniques may be intrusive or non-intrusive.

V.2. RISK BASED ASSESSMENT PERSONNEL

A multidisciplinary and experienced team is expected to be formed for the RBI assessment. The effectiveness of the RBI process is dependent upon the knowledge and experience of the team members and, as a minimum, the following disciplines are necessary inclusions:

— RBI facilitator certified to the requirements of Ref. [91];
— Materials and inspection specialist to identify and assess damage mechanisms and condition data to assess likelihood of damage mechanisms occurring;
— Maintainers and operations personnel of the organization;
— Technical processing representatives (e.g. hydrometallurgist).

Other disciplines may need to be involved, as required, depending on the type of equipment or site specific conditions.

An external consulting company will often be engaged to perform the RBI process as they will have access to the range of specialists required to complete an assessment (i.e. corrosion engineers, materials engineers, etc.). The identification of damage mechanisms and likelihood of them occurring would be completed by these specialists, however the organization's operational personnel will need to be included in the consequence assessment as they better understand the site specific consequences of the failure on the work force, environment, surrounding communities, reputation, production, revenue, etc.

V.3. RISK BASED ASSESSMENT METHODOLOGY

V.3.1. Planning

Risk based inspection planning is conducted at the equipment level (see Section 4.4 for information on equipment hierarchy), to assess the risk of failure and to utilize the risk outcome in inspection planning. The outcome is the RBI interval for the equipment item.

Equipment selected for risk based assessment may be identified by relevant site personnel or selected for its high cost of maintenance or high risk level, or where an improved understanding of damage mechanisms and controls to mitigate risk is needed. Ammonia vessels are a good example of where RBI is applied, due to the cost of inspection, risk associated with inspection itself and also the safety and environmental risks associated with these vessels while in operation.

V.3.2. Data and information collection

To conduct an RBI assessment, relevant physical, operational and maintenance inspection and repair data needs to be collected, collated and reviewed to ensure justification of decisions and consistency in decision making. Typical information collected includes:

— Equipment identification;
— Design data (e.g. design standards, pressures, temperatures, etc.);
— Relevant process flow diagrams, piping and instrumentation diagrams and general arrangement drawings;
— Operating and control philosophy (e.g. operating temperature/pressure, densities, fluid compositions, cyclic operation, vibration, etc);
— Control philosophy;

— Deviations from design (e.g. over temperature or over pressure);
— Materials of construction;
— Detail of internal and external coatings and insulation;
— Inspection history and results;
— Modification/repair history;
— Corrosion and erosion monitoring information.

V.3.3. Identifying damage mechanisms and failure modes

Guidance and general references for damage mechanisms in mineral processing plants can be found in the following publications:

— AS/NZS 3788:2006, Pressure Equipment – In-service Inspection [71];
— API RP 571:2020, Damage Mechanisms Affecting Fixed Equipment in the Refining Industry [92]. This publication identifies 171 possible damage mechanism, and although written in the context of the oil and gas industry, many of the damage mechanisms are also relevant to mineral processing.

The failure mode is dependent upon the degradation mechanism, equipment design and materials selection. Modes of failure describe how the equipment will fail as a result of the degradation mechanism, this typically includes weep, perforation, ductile rupture, brittle rupture and distortion. For example, excessive corrosion (i.e. the damage mechanism) can cause a pressure vessel to rupture (i.e. the failure mode). When performing an RBI assessment, the corrosion allowance stated in the manufacturers data report is used to determine the minimum acceptable wall thickness.

V.3.4. Assessing the likelihood of failure

The inspection history review and determination of the likelihood of failure is conducted by asset integrity specialists (i.e. a combination of inspectors and engineers).

Likelihood of failure is determined by the following factors:

— Identification of damage mechanisms and ranking their severity (e.g. the damage mechanism of corrosion will have a much higher likelihood for mild steel then for stainless steel).
— Review of inspection history to:
  - Determine what inspections have been conducted;
  - Determine the quality of those inspections;
  - Evaluate the inspection findings (i.e. was the expected damage mechanism found, and was its rate as expected? If the damage mechanism has been found in previous inspections, this is indicative that it will be found again in the future);
  - Review the operating history for the significance of any deviations from original design intent.

Once this information has been gathered an assessment can be made of the likelihood of failure for each damage mechanism using the operation's likelihood descriptors established by the organization (see Section 2.5.2). When using the RBI methodology, this likelihood of failure does not consider the following factors, as they are outside standard controls or design parameters, or cannot be controlled:

- — Seismic activity.
- — Weather extremes beyond that expected for the location.
- — Operation outside of the:
  - Original design intent;
  - Startup and shut down conditions (e.g. pressure, temperature and fluid chemistry).
- — Overpressure due to relief device failure.
- — Errors by operations personnel.
- — Design error.
- — Substitution of materials of construction.
- — Faulty installation.

V.3.5. Assessing the consequence of failure

The failure mode of a piece of equipment affects the severity of the consequences. For example, the consequences expected from a weep may be very different to the consequences expected from a brittle rupture. Hence, the consequence determination is conducted after the likelihood determination. The severity of the consequences of failure is ranked using the operation's descriptors established by the organization. If there is a range of consequence severities identified, the maximum reasonable consequence are to be used. Section 2.5.1 provides detailed information on both these subjects.

Site personnel, with representatives from the plant, operations, and maintenance are expected to be involved in determining the level of consequence for each impact type identified as a result of the damage mechanism.

V.3.6. Risk

The process of determining risk by combining the results of the likelihood and consequence is described in Section 2.5.

Once the risk ranking is determined, the damage mechanism contributing to the highest risk, and/or other unacceptable risks, can be identified and maintenance strategies developed including inspection type and frequency. Inspections can therefore be planned based on the damage mechanisms, failure rate and their corresponding risk.

The RBI process also identifies whether consequence or likelihood (or both) are driving risk. For likelihood to be driving overall risk, there would be a high likelihood damage mechanism with a low consequence. In this situation, where likelihood of failure is high (i.e. the damage mechanism is known, exists and is likely to occur), there is potential for risk management through inspection, particularly if the inspection scope was produced as part of the RBI assessment. Where consequence is driving the risk (i.e. low likelihood damage mechanism with high consequence), an increased inspection frequency will not reduce the likelihood of detecting the damage mechanism as there is already a low likelihood of it occurring.

V.4. RISK MANAGEMENT AND INSPECTION ACTIVITIES

Equipment inspection does not prevent damage from occurring, however it serves to identify failure rates and better plan maintenance repair strategies to reduce the risk of unplanned failure. Overall, an effective inspection programme can identify, monitor and measure deterioration and helps predict when the deterioration will reach a critical level.

Targeted inspection techniques will improve the ability to predict rates of deterioration, leading to less uncertainty regarding failure. Repair or replacement strategies can then be planned and implemented, prior to failure. These reductions in uncertainty lead directly to a decrease in risk.

An inspection methodology that is capable of reliably detecting the expected damage mechanisms needs to be used. Ideally, the methodology will reliably detect the expected damage at its lowest level or earliest stages. This, however, cannot always be achieved in practice. As a minimum, the inspection methodology needs to be capable of detecting the damage mechanism prior to it threatening integrity within the planned frequency of inspection.

An example of an RBI management and inspection strategy used within a uranium processing facility based on level of risk is shown in Table 31. The inspection intervals are provided as a guide only.

TABLE 31. RBI INTERNAL INSPECTION INTERVAL – TANKS AND PRESSURE EQUIPMENT

Risk band	**Action**	**Maximum tank inspection interval (years)**	**Maximum pressure vessel inspection interval (years)**	**Maximum boiler inspection interval (years)**
Critical	Unacceptable – Inspect within 3 months	0–0.25	0–0.25	0–0.25
High	Inspect at next shutdown or within interval, whichever is sooner	2	2	0.5
Moderate	Possible inspection deferral	8	4	1
Low	Inspection deferral likely	15	8	4
None	Extended deferral	25	16	8

A risk based inspection programme requires oversight by a competent person qualified for all aspects of the programme implementation. No methodology can foresee all possible situations and occasionally the output of the RBI assessment may need to be modified to accommodate site specific conditions. In some cases, other mechanisms may provide reliable methods for determining inspection intervals. For example, an active, predictable wall loss mechanism may be identified and an inspection interval, based upon expected thickness at the next inspection, may be used, rather than the intervals provided in the RBI programme (i.e. Table 31).

V.5. RISK BASED INSPECTION ASSESSMENT REVIEW

The RBI assessments for critical equipment are expected to be reviewed in the following circumstances:

- Following process excursions outside the original design intent;
- After a set time period. Maximum extension intervals are provided in Ref. [71];
- Before and after planned maintenance, shutdowns and inspections.

APPENDIX VI.

TANK INSPECTION GUIDANCE

The inspection guidance for above ground storage tanks provided in this appendix, is to be read in conjunction with Section 6.5, which provides a general introduction to tank inspections. Examples of national standards are provided only to illustrate and provide further reading on the types of industry guidelines that support tank inspections.

When planning for inspections, the damage mechanisms specific to the equipment need to be considered, as this will inform the types of inspections and tests that will be performed. Additionally, inspections are expected to be performed by personnel with the relevant qualifications, competencies and experience.

VI.1. SHELL

The integrity of the shell thickness needs to be determined by using external UT measurements in accordance with applicable standards, for example Ref. [93]. In cone bottom tanks (e.g. thickeners), the shell thickness inspection is also applied to the cone section.

The vertical portion of the tank, which includes the full height of the tank, is assessed at the four compass points using an open grid with appropriate grid spacing. A similar grid spacing methodology is also employed on the cone section.

The following additional checks are made on cone bottom tanks:

- — Check thickness of support legs within the bottom 100 mm of the section;
- — Examine support brace/legs to assure design compliance;
- — Examine the earthquake ring and wind girder for signs of buckling, plastic deformation or corrosion;
- — Check hold down bolts for integrity.

If corrosion is detected on the shell of a tank, then those corroded areas that may adversely affect the performance or structural integrity of the tank will be retested at the deteriorated zone using a closed grid with appropriate grid spacing. A contour map is then used to permit an engineering assessment of integrity in accordance with a relevant standard, such as Ref. [42].

VI.2. SHELL WELDS

The wall welds are to be inspected using MPI, DPI or PA inspection, both internally and externally.

VI.3. MAINTENANCE HOLE AND NOZZLE WELDS

It is good industry practice that all welds on maintenance holes and nozzles, both internal and external, are inspected. However, for low risk rated tanks it may be sufficient to only perform visual inspections.

The opening to shell welds and the opening to reinforcing plate welds (where applicable) are to be inspected and tested using a relevant standard. Examples of standards, for specific testing techniques, are provided in Refs [41, 51, 55, 94].

VI.4. NOZZLE SHELL THICKNESS

The thickness of nozzle walls are to be inspected for thinning and pitting using a minimum of six UT measurements per nozzle. On large bore nozzles (300 mm diameter and greater) additional inspection points are to be used.

Ultrasonic thickness testing methods are to be performed in accordance with applicable standards, for example Ref. [93].

VI.5. EXTERNAL VISUAL

The external areas of tanks are to be visually inspected for signs of excessive loss of protective coating, general corrosion, localized erosion, erosion adjacent to nozzles, holes, pitting, deformation of walls, cracking and structural damage.

Particular attention are to be given to projection plate condition, plinth seal and the critical zone of the tank, as well as earthquake rings and strengthening ribs where process spillage can accumulate and result in accelerated corrosion or unacceptable loading surcharges. Any new and active corrosion are to be marked for further NDT.

For flat bottom tanks on the ground located within secondary containment areas, attention needs to be given to any soil, moisture, process materials and vegetation that may accumulate against annular rings and walls, causing accelerated corrosion. Any tank subsidence is also to be noted and reported. For insulated tanks, the inspection needs to pay particular attention to the condition of lagging and cladding.

Defects are to be further investigated and assessed in accordance with an appropriate standard and further NDT is to be conducted where required to assess the degree of damage.

A detailed tank inspection checklist for inspecting tanks can be found in appendix V of Ref. [42], which is used by tank inspectors to guide visual inspection.

VI.6. TANK LID/ROOF

Tanks with a lids or roof have the potential for the roof/lid to be used as pedestrian walkways to access entry hatches, mechanical equipment or tank control instrumentation. The roof area may also support equipment used for tank cleaning or maintenance refurbishments.

An assessment of lid/roof integrity, at a minimum, contains the following:

- An accurate drawing of the tank roof, including an outline of each plate to ensure that each roof plate can be uniquely identified.
- A SLOFEC™ technique, or similar, that uses the eddy current principle in combination with a magnetic field as a method for detecting loss of metal thickness and corrosion mapping due to underroof corrosion.
- Ultrasonic thickness measurements, as follows:
  - Rectangular plates: Six readings per plate;
  - Sketch plates (those cut to a rounded shape): Three readings per plate;
  - Roof near the shell junction: Ten readings at 150 mm centres at four locations on the perimeter;

- Plate laps: Ten readings at 150 mm centres at four randomly chosen locations on roof.

Ultrasonic thickness testing methods are to be performed in accordance with an applicable standard, for example Ref. [93].

VI.7. INTERNAL FLOOR PLATE (FLAT BOTTOM TANKS)

The integrity of internal tank floor plates is evaluated by measuring the plate thickness. Several techniques can be employed for the thickness measurements. SLOFEC™ and MFL are the preferred in situ screening techniques of tank floors. The SLOFEC™ technique is an efficient method for detecting loss of metal thickness due to under floor corrosion. Where SLOFEC™ cannot be used, the floor thickness is to be measured on a grid pattern of six readings per plate. The annular ring thickness is measured at 20 equally spaced locations around the tank circumference. In addition:

— The floor welds and the wall to floor weld is be inspected using MPI. In addition, the floor welds and wall to floor weld are also inspected using dye penetrant inspection.
— It is good industry practice that 50% of general floor welds are inspected. 100% of the floor to wall welds are inspected.

Fluorescent MPI may be used to detect smaller imperfections.

Further guidance is provided in Refs [55, 94].

VI.8. CRITICAL ZONE

The critical zone of the tank comprises the area extending within 80 mm inward, outward and upward from the floor-to-shell weld, which is not accessible by either SLOFEC™, MFL, or RMS inspection techniques, and is commonly referred to as the 'dead zone' for this type of inspection. This area is to be inspected using in situ manual UT testing. The critical zone integrity, as suggested by the name, is the most critical area for the tank's structural integrity and therefore it needs to be a priority of any inspection to verify that it is in an acceptable condition, namely:

— The floor thickness in the critical zone remains above the minimum allowable thickness;
— The floor-to-shell butt weld has no cracks.

Pulsed eddy current may be employed to screen the condition of the projection plate (the floor plate protruding outside the shell) and the floor side of the critical zone. More details on projection plate inspection is provided in Section VI.9. Pulsed eddy current is the only tool that is currently available on the market to carry out in-service inspection of the critical zone of above ground storage tanks, where access inside the tank is restricted. The probe uses a highly durable 4.8 mm thick titanium blade but requires a gap between the projection plate and the tank's foundation to allow the 400 mm deep insertion to acquire a reading. As a screening tool, any defect indications will need to be proved using manual UT, which may require access from inside the tank.

In addition, regular vertical banding checks are to be carried out around the critical zone to monitor for any buckling or distortion, in accordance with a relevant standard, such as section 10.5.5 in Ref. [42].

VI.9. PROJECTION PLATE

The thickness and condition of the projection plate in the critical zone of the tank can be measured manually using UT (of the protrusion plate thickness), MPI (for the floor-to-shell butt weld integrity) or pulsed eddy current providing the probe can be inserted under the projection plate which requires the plinth seal removal. The width of the projection plate can be measured using a standard ruler.

VI.10. INTERNAL LINING

Tanks fitted with internal rubber lining can be visually inspected and, where possible, inspected using spark testing techniques to determine presence of any holes in the lining. The spark tester needs to be tuned to penetrate a certain thickness of the lining. Therefore, any significant loss or gouge from the lining will show up as a 'hole'. As small holes are difficult or impossible to visually see in rubber, all areas which indicate a 'hole' through the spark testing technique are to be repaired with a rubber patch.

Tanks fitted with internal rubber lining is to be further inspected in accordance with an applicable standard, such as Ref. [95]. More details on types of rubber linings used within a uranium processing facility, their application and quality controls are outlined in Section 6.9.

VI.11. INTERNAL VISUAL

The internal shell and process components of storage tanks is to be visually inspected for signs of excessive general corrosion, localized erosion, erosion adjacent to nozzles, holes, pitting, deformation of walls or floors, baffles, cracking and structural damage of columns and roof rafters where roofs are fitted.

VI.12. FLAT BOTTOM TANKS

For flat bottom tanks, the hold down bolts and concrete base which the tanks sit on are to be checked for integrity.

VI.13. HORIZONTAL FUEL STORAGE TANKS

The inspection regime for the tanks classified as horizontal fuel tanks includes:

- — Four equidistant circumferential scans to determine thickness of the vessel;
- — Visual inspection.

APPENDIX VII.

CRITICAL RUBBER LINED TANK DAMAGE MECHANISM AND FAILURE MODES

An example of damage mechanisms and failure modes for rubber lined mild steel tanks is shown in Table 32. Damage mechanisms will be unique to each processing facility due to site specific conditions.

TABLE 32. RUBBER LINED TANK DAMAGE MECHANISMS AND FAILURE MODES

Damage mechanism	Failure mode	Impact description	Controls	Failure mode with controls	Risk comments
			Location: Tank Shell		
Corrosion due to bubble behind rubber. Liquid behind rubber lining very slowly corrodes the shell	Leak	**Production:** Tank taken offline and dropped below leak and repaired. Tank offline for 2–3 months for repair **Environment:** NA if bermed/bunded	Four yearly (or as required) internal visual inspection to remove and repair large bubbles. Leak response: drain tank and repair leak.	Leak	Contents of bubbles have been tested. Composition is close to water (4<pH<8, slurry pH<2.5) and have very slow corrosion rates. If bubble areas do corrode through the shell, they create a leak which is picked up by production operators
		Health and Safety: Slurry in contact with person in area. *See Risk Comments.	PPE, area inductions, limited access	Leak	*Health and safety risk associated with a leak are the same for other damage mechanisms with similar failure modes and are therefore not repeated for each damage mechanism
Corrosion due to a bonding defect at installation. Rubber joint defect allows acidic solution to corrode steel leading to vertical defect causing collapse	Rupture	**Compliance:** Site shut down	Rubber installation quality assurance (including temperature, humidity, surface preparation, steel chlorides, warm up rates for cold weather)	Leak	Under normal circumstances rubber does not pull away if installed correctly. This damage mechanism applies particularly to newly applied rubber patches where the vertical length exceeds the critical vertical defect length
Corrosion due to general rubber degradation (wear, chemical attack, high temperature)	Leak	**Production:** Tank taken offline and dropped below leak and repaired. Tank offline for 2–3 months for repair	Four yearly internal inspection. Inspection intent is to provide information on rubber integrity.	Leak	Rubber lining is resistant to wet material wear and chemical attack. Minimal evidence of this occurring. Chemical attack from organics in system needs to be understood and controlled

TABLE 32. RUBBER LINED TANK DAMAGE MECHANISMS AND FAILURE MODES (cont.)

Damage mechanism	Failure mode	Impact description	Controls	Failure mode with controls	Risk comments
			Location: Tank Shell		
Corrosion of shell due to rubber lining failure caused by impact damage from agitator blade failure	Leak	**Production:** Tank taken offline and dropped below leak and repaired. Tank offline for 2–3 months	Improved quality assurance of blade attachment welding for new agitators including weld specification and NDT	Leak	Failed blades from agitator have blades fall to the floor beneath agitator on to the solids bed
Corrosion of the shell due to rubber damage from swinging baffle arm. Baffle arm fails due to fatigue cracking of arm leading to vertical defect causing collapse	Rupture	**Compliance:** Site shut down	Baffle supports to withstand fatigue cracking. Additional supports on each baffle between lower support arms	Rupture	Baffle arm failure to consider both fatigue cracking and rubber lining failure
Corrosion due to foreign object dropped into tank damaging rubber lining	Leak	**Production:** Tank taken offline and dropped below leak and repaired. Tank offline for 2–3 months for repair	Reporting of foreign objects dropped into tank. Mitigating control will be on case-by-case basis. Worst case scenario for large object would be to drop tank level to retrieve object and inspect tank walls	Leak	Lining resistant to impacts from small objects. Very large object would need to be dropped to cause hole in rubber. Would not lead to catastrophic failure and rather hole in shell and leak due to rubber damage
General corrosion of external tank due to lack of paint maintenance	Leak	**Production:** Tank taken offline and dropped below leak and repaired. Tank offline for 2–3 months	Monthly operator external inspection. External NDT and tank inspections	Leak	
Corrosion of external tank due to spillage	Rupture	**Compliance:** Site shut down	Operator housekeeping initiatives implemented to ensure spillage is cleaned up within the shift. Daily operator inspections. Monthly operator external inspection	Leak	Removing product build up is administrative and reduces the likelihood only. Plant cleanliness and reduced spillage relies on consistent and high quality production practices
Corrosion of shell due to rubber lining damage due to lowering sump pump into tank during pump out	Rupture	**Compliance:** Site shut down	Drain tank using drain valve and piping/pimp arrangement Change pump to submersible pump with no ability to mechanically damage rubber lining for when drain pump blocks and tank needs to be pumped out from the surface	Leak	

TABLE 32. RUBBER LINED TANK DAMAGE MECHANISMS AND FAILURE MODES (cont.)

Damage mechanism	Failure mode	Impact description	Controls	Failure mode with controls	Risk comments
Location: Tank Shell and Floor					
Corrosion due to damaged rubber lining from maintenance works (i.e. removal of dried scale)	Leak	**Production:** Tank taken offline and dropped below leak and repaired. Tank offline for 2–3 months	Ensure maintenance activities include rubber integrity considerations. This includes spark testing of floor and visual inspection of walls to ensure no damage to rubber prior to exiting tank	Leak	Wet blast can be used for semi hardened scale. Avoid use of scaling bars or jack hammers due to high potential of damaging the lining
Corrosion of shell due to rubber lining damage from failure of the agitator shaft (i.e. comes into contact with the shell and damages rubber lining)	Leak	**Production:** Tank taken offline and dropped below leak and repaired. 1 tank off for 2–3 months	New agitator quality assurance and NDT requirements specified for new and refurbished agitators Response plan in place to check and inspect agitator following failure. Refurbished agitator quality control prior to installation (VI, DPI, UT inspection).	Leak	
Location: Tank Floor					
Under floor corrosion due to water or slurry ingress	Rupture	**Compliance:** Site shutdown	Monthly operator visual inspection of seal. External inspection for water ingress under protrusion plate. Internal inspection using SLOFEC™ scanning on rubber surface.	Rupture	If the mastic seal is found to be deteriorated during operator visual inspection it is reported immediately and repaired
Corrosion of floor due to a bonding defect at installation	Leak	**Production:** Minimal impact if scale bed exists. Scale build up would prevent leak	Rubber lining quality assurance. Internal inspection allows inspection of floor and cleaning	Leak	

APPENDIX VIII.

ASSET MANAGEMENT RELATED INTERNATIONAL GUIDANCE AND STANDARDS

This appendix provides a summary, shown in Table 33, of relevant publications for further guidance on asset management of uranium mining and processing facilities. The table includes a section reference within this publication and high level information about the scope of application in asset management.

TABLE 33. LIST OF ASSET MANAGAMENT RELATED PUBLICATIONS

Related publication	Section reference and scope
American Petroleum International – API Standards	
API 570: Piping Inspection Code: In-service Inspection, Repair, Rating, and Alteration of Piping Systems	**Section 6.7 Piping.** This publication provides inspection, rating, repair, and alteration procedures for metallic and fibreglass reinforced plastic piping systems and their associated pressure relieving devices
API RP 571: Damage Mechanism Affecting Fixed Equipment in the Refining Industry	**Appendix V. Risk Based Assessment Process.** This publication provides a list of all damage mechanisms affecting fixed equipment and refining equipment, including details of the 171 possible damage mechanisms. Although written in the context of the oil and gas industry, many of the damage mechanisms are also relevant to minerals processing
API RP 576: Inspection of Pressure-Relieving Devices	**Section 6.8 Pressure Safety and Pressure Relief Valves.** This publication provides further guidance on determining intervals for pressure safety valve testing and overhaul
API RP 580: Risk Based Inspection	**Appendix V. Risk Based Assessment Process.** This publication provides the basic guidelines for implementing and maintaining a risk based inspection programme, and guidelines for assessing an inspection programme
API 620: Design and Construction of Large, Welded, Low Pressure Storage Tanks	**Section 6.5 Storage Tanks.** This publication provides design and construction of large field-assembled, welded, low pressure carbon steel above ground storage tanks
API STD 650: Welded Steel Tanks for Oil Storage	**Section 6.5 Storage Tanks.** This publication established minimum requirements for material, design, fabrication, erection, and inspection for storage tanks
API STD 653: Tank Inspections, Repair, Alteration and Reconstruction	**Section 6.1 Condition Based Monitoring, Non-destructive Testing, 6.5 Storage Tanks.** This publication provides the minimum requirements for maintaining the integrity of tanks
Standards Australia / Standards New Zealand – AS/NZS Standards	
AS 1200: Pressure Equipment	**Section 6.6 Pressure Vessels.** The publication provides basic requirements and good practice for the design, materials, manufacture, examination, testing, installation, conformity assessment, commissioning, operation, inspection, maintenance, repair, alteration and disposal of pressure equipment
AS 1210: Pressure Vessels	**Section 6.6 Pressure Vessels.** This publication provides the minimum requirements for the materials, design, manufacture, testing, inspection, certification, documentation and dispatch of fired and unfired pressure vessels
AS 1271: Safety Valves, Other Valves, Liquid Level Gauges, and Other Fittings for Boilers and Unfired Pressure Vessels	**Section 6.8 Pressure Safety and Pressure Relief Valves.** This publication provides requirements for the design, construction and testing of safety valves, and others, for use on boilers and unfired pressure vessels and their associated piping

TABLE 33. LIST OF ASSET MANAGAMENT RELATED PUBLICATIONS (cont.)

Related publication	Section reference and scope
Standards Australia / Standards New Zealand – AS/NZS Standards	
AS 1418.1: Cranes, Hoists and Winches, Part 1: General Requirements	**Section 6.3 Cranes and Lifting Equipment.** This series of publications provide requirements for cranes, hoists, winches and the associated components and appliances
AS/NZS 1554.5: Structural Steel Welding – Part 5: Welding of Steel Structures Subject to High Levels of Fatigue Loading	**Appendix VI. Tank Inspection Guidance.** This publication specifies requirements for the welding of steel structures made up of combinations of steel plates, sheet or sections
AS/NZS 1554.6: Structural Steel Welding – Part 6: Welding Stainless Steels for Structural Purposes	**Appendix VI. Tank Inspection Guidance.** This publication specifies requirements for the welding of steel structures made up of combinations of steel plates, sheet or sections
AS/NZS 1660: Test Methods for Electrical Cables, Cords and Conductors	**Section 6.1 Condition Based Monitoring, Non-destructive Testing.** This series of publications sets out the electrical test for components parts and complete cables, cords and conductors
AS 2452.3: Non-Destructive Testing – Determination of Thickness Part 3 – Use of Ultrasonic Testing	**Appendix VI. Tank Inspection Guidance.** The publication specifies methods for determination of thickness of material based on the use of UT testing
AS 2593: Boilers – Safety Management and Supervisory Systems	**Section 6.8 Pressure Safety and Pressure Relief Valves.** Section 4.7 of this publication provides further guidance on determining intervals for in service inspections of pressure relief devices
AS 2550.15: Cranes, Hoists and Winches – Safe Use, Part 15: Concrete Placing Equipment	**Section 6.3 Cranes and Lifting Equipment.** This publication provides the requirements for safe use of concrete placing for cranes, hoist and winches
AS/NSZ 3788: Pressure equipment – In-service Inspection	**Sections 6.6, 6.7 & 6.8 Pressure Equipment (Various).** This publication specifies the minimum requirements for the inspection, repair and alteration of in-service boilers, pressure vessels, and safety equipment, associated controls and execution of such activities
AS 3873: Pressure Equipment – Operation and Maintenance	**Section 6.6 Pressure Vessels.** This publication provides the minimum requirements and guidance on the operation and maintenance of boilers, pressure vessels and associated control and safety equipment
AS 4041: Pressure Piping	**Section 6.7 Piping.** This publication provides the minimum requirements for the materials, design, fabrication, testing, inspection, reports and pre-commissioning of piping subject to pressure
AS 4343: Pressure Equipment – Hazard Levels	**Section 6.6 & 6.7 Pressure Equipment (Various).** This publication specifies criteria for determining the hazard levels of various types of pressure equipment
AS 4971: Inspection and Integrity Monitoring of Large Steel Vertical Petroleum Storage Tanks	**Section 6.5 Storage Tanks.** The publication provides requirements for tank inspection and tank monitoring for ensuring structural integrity of petroleum storage tanks
American Society of Mechanical Engineers – ASME Standards	
ASME B31.3. Process Piping	**Section 6.7 Piping.** The publication covers materials and components, design, fabrication, assembly, erection, examination, inspection, and testing of piping
ASME BPVC, Section V – Nondestructive Examination	**Appendix VI. Tank Inspection Guidance**. This publication provides requirements and methods for NDT of boiler and pressure vessels
ASME BPVC, Section VIII – Rules for Construction of Pressure Vessels Division 1	**Section 6.6 Pressure Vessels.** The publication provides requirements applicable to the design, fabrication, inspection, testing and certification of pressure vessels

TABLE 33. LIST OF ASSET MANAGAMENT RELATED PUBLICATIONS (cont.)

Related publication	Section reference and scope
American Society for Testing Materials – ASTM Standards	
ASTM D7669: Standard Guide for Practical Lubricant Condition Data Trend Analysis	**Section 6.1 Condition Monitoring Techniques, Non-destructive Testing.** This publication provides further guidance on for oil analysis on mining and processing equipment
ASTM E507: Standard Practice for Flux Leakage Examination of Ferromagnetic Steel Tubular Products	**Section 6.1 Condition Based Monitoring, Non-destructive Testing.** This publication covers the application and standardization of equipment using flux leak test methods
ASTM E709: Standard Guide for Magnetic Particle Testing	**Section 6.1 Condition Based Monitoring, Non-destructive Testing.** This publication covers techniques for MPI
ASTM E1032: Standard Practice for Radiographic Examination of Weldments Using Industrial X Ray Film	**Section 6.1 Condition Based Monitoring, Non-destructive Testing.** The publication provides a uniform procedure for radiographic examination of welding instruments using industrial radiographic film
ASTM E3052: Standard Practice for Examination of Carbon Steel Welds Using an Eddy Current Array	**Section 6.1 Condition Based Monitoring, Non-destructive Testing.** This publication covers the use of an eddy current array or sensor for NDT
British Standards Institution – BS Standards	
BS 6374. Lining of Equipment with Polymetric Materials for the Process Industries	**Appendix VI. Tank Inspection Guidance.** This series of publications specifies requirements for the lining of equipment using thermoplastics
BS 7159. Code for Practice for Design and Construction of Glass-reinforced Plastics (GRP) Piping Systems for Individual Plants of Sites	**Section 6.1 Condition Based Monitoring, Non-destructive Testing.** This publication provides guidance for the design and construction of glass reinforced plastics
BS EN 31010: Risk Management – Risk Assessment Techniques	**Section 2.5 Risk Assessment.** This publication gives guidance on selecting and applying risk assessment techniques
Canadian Standards Association – CSA Standards	
CSA B51. Boiler, Pressure Vessel and Pressure Piping Code	**Section 6.6 Pressure Vessels.** This publication contains requirements for boilers, pressure vessels, pressure piping and fittings. It promotes safe design, construction, installation, operation, inspection, testing, and repair practices
The Engineering Equipment and Materials Users Association – EEMUA Standards	
EEMUA Publication 159: Above Ground Flat Bottomed Storage Tanks – A Guide to Inspection, Maintenance and Repair	**Section 6.5 Storage Tanks.** This publication established the essential inspection and maintenance requirements for in-service storage tanks
European Committee for Standardization – EN Standards	
EN 583: Non-destructive testing – Ultrasonic Examination	**Section 6.1 Condition Based Monitoring, Non-destructive Testing.** This series of publications defines the principals for UT examination of industrial products that permit transmission of ultrasound
EN 1330-9: Non-Destructive Testing – Terminology – Part 9: Terms Used in Acoustic Emission Testing	**Section 6.1 Condition Based Monitoring, Non-destructive Testing.** This publication provides NDT terminology used in acoustic emission testing
EN 13554: Non-destructive testing – Acoustic Emission Testing. General principals	**Section 6.1 Condition Based Monitoring, Non-destructive Testing.** This publication specifies methods and guidelines for NDT using active thermography with laser excitation
EN 14127: Non-destructive Testing – Ultrasonic Thickness Measurement	**Section 6.1 Condition Based Monitoring, Non-destructive Testing.** This publication specifies the principals for UT measurement

TABLE 33. LIST OF ASSET MANAGAMENT RELATED PUBLICATIONS (cont.)

Related publication	Section reference and scope
European Committee for Standardization - EN Standards	
EN 14584: Non-destructive Testing – Acoustic emission Testing – Examination of Metallic Pressure Equipment During Proof Testing – Planar Location of AE Sources	**Section 6.1 Condition Based Monitoring, Non-destructive Testing.** This publication describes the method for conducting acoustic emission testing of metallic pressure equipment using planar location
EN 15341: Maintenance Key Performance Indicators	**Section 5.3 Performance Management and Continuous Improvement.** This publication lists key performance indicators of the maintenance function and provides guidelines to define indicators for continuous improvement
EN 16018: Non-destructive Testing – Terminology – Terms Used in Ultrasonic Testing with Phased Arrays	**Section 6.1 Condition Based Monitoring, Non-destructive Testing.** This publication provides terminology for ultrasonic testing with phased arrays
International Organization for Standardization – ISO Standards	
ISO 3058: Non-destructive testing – Aids to Visual Inspection – Selection of Low Power Magnifiers	**Section 6.1 Condition Based Monitoring, Non-destructive Testing.** This publication specifies the characteristics of low power magnifiers for the inspection of surfaces
ISO 3452: Non-destructive Testing – Penetrant Testing	**Section 6.1 Condition Based Monitoring, Non-destructive Testing.** The publication series specifies a method of penetrant testing use to detect discontinuities
ISO 5579: Non-destructive Testing – Radiographic Testing of Metallic Materials Using File and X- or Gamma Rays – Basic Rules	**Section 6.1 Condition Based Monitoring, Non-destructive Testing.** This publication outlines the general rules for industrial X- and gamma-radiography for flaw detection purposes
ISO 9934: Non-destructive Testing – Magnetic Particle Testing	**Section 6.1 Condition Based Monitoring, Non-destructive Testing.** This series of publications provides further guidance for the MPI of ferromagnetic materials
ISO 13588: Non-destructive Testing of Welds – Ultrasonic Testing – Use of Automated Phased Array Technology	**Section 6.1 Condition Based Monitoring, Non-destructive Testing.** This standard specifies the application of phased array technology for UT testing of fusion-welded joints
ISO 14224: Petroleum and Natural Gas Industries – Collection and Exchange of Reliability and Maintenance Data for Equipment	**Section 4, 5 and 6 (Various).** The publication provides a comprehensive basis for collection of reliability and maintenance data during the operational life cycle of equipment
ISO 15548: Non-destructive Testing – Equipment for Eddy Current Examination	**Section 6.1 Condition Based Monitoring, Non-destructive Testing.** This series of publications identifies the functional characterization of eddy current instruments and methods for measurement and verification
ISO 15549: Non-destructive Testing – Eddy Current Testing – General Principles	**Section 6.1 Condition Based Monitoring, Non-destructive Testing.** This publication defines the principals to be applied to eddy current examination of products and materials
ISO 16809: Non-destructive Testing – Ultrasonic Thickness Measurement	**Section 6.1 Condition Based Monitoring, Non-destructive Testing.** This standard specifies the principals for UT measurement
ISO 17640: Non-destructive Testing of Welds – Ultrasonic Testing – Techniques, Testing Levels and Assessments	**Section 6.1 Condition Based Monitoring, Non-destructive Testing.** This standard specified the techniques for the manual UT testing of fusion-welded joints
ISO 17637: Non-destructive Testing of Welds – Visual Testing of Fusion-welding Joints	**Section 6.1 Condition Based Monitoring, Non-destructive Testing.** This standard specifies the visual testing of fusion welds in metallic materials
ISO 18251: Non-destructive Testing – Infrared Thermography	**Section 6.1 Condition Monitoring Techniques, Thermography.** This series of publications provides further guidance on the infrared imaging systems and related equipment used on NDT

TABLE 33. LIST OF ASSET MANAGAMENT RELATED PUBLICATIONS (cont.)

Related publication	Section reference and scope
International Organization for Standardization – ISO Standards	
ISO 18436-4: Condition Monitoring and Diagnostics of Machines – Requirements for Qualification and Assessment of Personnel – Part 4: Field Lubrication Analysis	**Section 6.1 Condition Monitoring Techniques, Oil Analysis.** This publication provides further guidance on the requirements for qualification and assessment of personnel who perform machinery condition monitoring using field lubricant analysis
ISO 18436-7: Condition Monitoring and Diagnostics of Machines — Requirements for Qualification and Assessment of Personnel — Part 7: Thermography	**Section 6.1 Condition Monitoring Techniques, Oil Analysis.** The publication specifies the requirements for qualification and assessment of personnel who perform machinery condition monitoring and diagnostics using infrared thermography
ISO 20816: Mechanical Vibration – Measurement and Evaluation of Machine Vibration	**Section 6.1 Condition Monitoring Techniques, Vibration Analysis.** This series of publications provides further guidance on machine vibration measurement on rotating, non-rotating and non-reciprocating parts
ISO 20339: Non-destructive Testing – Equipment for Eddy Current Examination – Array Probe Characteristics and Verification	**Section 6.1 Condition Based Monitoring, Non-destructive Testing.** The publication identified the functional characteristics of eddy current array probes and methods for their measurement and verification
ISO 20669: Non-destructive Testing – Pulsed Eddy Current Testing of Ferromagnetic Metallic Components	**Section 6.1 Condition Based Monitoring, Non-destructive Testing.** The publication specifies the pulsed eddy current testing technique to perform thickness measurement
ISO 20769: Non-destructive Testing – Radiographic Inspection of Corrosion and Deposits in Pipes by X- and Gamma Rays	**Section 6.1 Condition Based Monitoring, Non-destructive Testing.** This series of publications specifies techniques of film and digital radiography
ISO 23277: Non-destructive Testing of Welds – Penetrant Testing – Acceptance Levels	**Section 6.1 Condition Based Monitoring, Non-destructive Testing.** The publication specified acceptance levels for indication from surface breaking imperfections in welds detected by penetrant testing
ISO 55000: Asset Management – Overview Principles and Terminology	**Section 3.1 Asset Management Framework.** This publication provides and overview of asset management and the benefits of adopting asset management
ISO 55001: Asset Management – Management Systems – Requirements	**Section 3.1 Asset Management Framework.** This publication specifies requirements for an asset management system withing the context of an organization
ISO 55002: Asset Management – Management Systems – Guidelines for the application of ISO 55001	**Section 3.1 Asset Management Framework.** This publication provides the guidelines for the application of an asset management system, in accordance with ISO 55001
South African Bureau of Standards – SABS Standards	
SANS 141: Glass-reinforced Polyester Laminates	**Section 6.10.3 Glass fibre reinforcement and finishing tissue.** This publication provides guidance for the design and construction of glass reinforced plastics.

REFERENCES

[1] OECD NUCLEAR ENERGY AGENCY, INTERNATIONAL ATOMIC ENERGY AGENCY, Uranium 2022: Resources, Production and Demand, OECD Publishing, Paris (2023).

[2] INTERNATIONAL ATOMIC ENERGY AGENCY, Leadership and Management for Safety, IAEA Safety Standards Series No. GSR Part 2, IAEA, Vienna (2016), https://doi.org/10.61092/iaea.cq1k-j5z3.

[3] NATIONAL OCCUPATION HEALTH AND SAFETY COMMISSION, Control of Major Hazard Facilities, National Standard NOHSC:1014, Australian Government Publishing Services, Canberra (2002).

[4] CHEMICAL INSPECTION AND REGULATION SERVICE, Regulations of the Safe Management of Hazardous Chemicals in China, Decree 591 of the State Council in China (2011), English Translation available from http://www.cirs-reach.com/China_Chemical_Regulation/Regulations_on_safe_management_on_hazardous_chemicals_China_2011.pdf.

[5] EUROPEAN PARLIAMENT AND COUNCIL OF THE EUROPEAN UNION, Directive of the 2012/18/EU European and of the Council on the Control of Major-accident Hazards Involving Dangerous Substances, Official Journal of the European Union (2012), http:// eur-lex.europa.eu/legal-content/EN/TXT/?uri=CELEX:32012L0018.

[6] THE HIGH PRESSURE GAS SAFETY INSTITUTE OF JAPAN, Overview of the High Pressure Gas Safety Act in Japan, The High Pressure Gas Safety Institute of Japan (KHK), Toyko 2019, https://www.khk.or.jp/english/overview_of_hpg_safety_act.html.

[7] UNITED STATES DEPARTMENT OF LABOUR, Process Safety Management of Highly Hazardous Chemicals, 29 CFR 1910.119, Occupational Safety and Health Administration, Washington (1992), https://www.osha.gov/process-safety-management/standards.

[8] RÖSSING URANIUM LIMITED, Country Focus: Namibia, Rössing - the great survivor (2003), https://www.rossing-com.info/reports/African_Mining_Jan_Feb_2003.pdf.

[9] WORLD INFORMATION SERVICE ON ENERGY, Issues at Operating Uranium Mines and Mills — Olympic Dam, Australia (2023), https://www.wise-uranium.org/umopauod.html.

[10] THE SYDNEY MORNING HERALD, WMCR drops $50m on Olympic Dam plant, Australia (2023), https://www.smh.com.au/business/wmcr-drops-50m-on-olympic-dam-plant-20030115-gdg3xu.html.

[11] INTERNATIONAL ATOMIC ENERGY AGENCY, Managing Organizational Change in Nuclear Organizations, IAEA Nuclear Energy Series No. NG-T-1.1, IAEA, Vienna (2014).

[12] INTERNATIONAL ATOMIC ENERGY AGENCY, Modifications to Nuclear Power Plants, IAEA Safety Standards Series No. SSG-71, IAEA, Vienna (2022).

[13] INTERNATIONAL ORGANIZATION FOR STANDARDIZATION, Asset Management: Management Systems — Requirements, ISO 55001:2024, ISO, Geneva (2024).

[14] INTERNATIONAL ORGANIZATION FOR STANDARDIZATION, Asset Management: Management Systems — Guidelines for the application of ISO 55001, ISO 55002:2018, ISO, Geneva (2018).

[15] BRITISH STANDARDS INSTITUTION, Risk Management: Risk Assessment Techniques, BS EN 31010:2019, BSI, London (2019).

[16] HEALTH AND SAFETY EXECUTIVE, Risk Criteria for Land-use Planning in the Vicinity of Major Industrial Hazards, H.M. Stationery Office, Great Britain (1989).
[17] HEALTH AND SAFETY EXECUTIVE, 5 Steps to Risk Assessment, INDG163, 1st rev., Health and Safety Executive, United Kingdom (1994).
[18] CENTRE FOR CHEMICAL PROCESS SAFETY, Guidelines for Chemical Process Quantitative Risk Analysis, 2nd edn., John Wiley & Sons, Inc., (1999).
[19] COMMITTEE FOR THE PREVENTION OF DISASTERS, Guidelines for Quantitative Risk Assessment 'Purple Book', 3rd edn., Gevaatlijke Stoffen, Netherlands (2005).
[20] INTERNATIONAL ATOMIC ENERGY AGENCY, Occupational Radiation Protection, IAEA Safety Standards Series No. GSG-7, IAEA, Vienna (2018).
[21] INTERNATIONAL ATOMIC ENERGY AGENCY, Occupational Radiation Protection in the Uranium Mining and Processing Industry, IAEA Safety Reports Series No. 100, IAEA, Vienna (2020).
[22] INTERNATIONAL ATOMIC ENERGY AGENCY, Radiation Protection of Itinerant Workers, IAEA Safety Reports Series No. 84, IAEA, Vienna (2015).
[23] EUROPEAN COMMISSION, FOOD AND AGRICULTURE ORGANIZATION OF THE UNITED NATIONS, INTERNATIONAL ATOMIC ENERGY AGENCY, INTERNATIONAL LABOUR ORGANIZATION, OECD NUCLEAR ENERGY AGENCY, PAN AMERICAN HEALTH ORGANIZATION, UNITED NATIONS ENVIRONMENT PROGRAMME, WORLD HEALTH ORGANIZATION, Radiation Protection and Safety of Radiation Sources: International Basic Safety Standards, IAEA Safety Standards Series No. GSR Part 3, IAEA, Vienna (2014), https://doi.org/10.61092/iaea.u2pu-60vm.
[24] INTERNATIONAL ATOMIC ENERGY AGENCY, Milestones in the Development of a National Infrastructure for the Uranium Production Cycle, IAEA Nuclear Energy Series No. NF-G-1.1, IAEA, Vienna (2023).
[25] INTERNATIONAL ATOMIC ENERGY AGENCY, Application of the Management System for Facilities and Activities, IAEA Safety Standards Series No. GS-G-3.1, IAEA, Vienna (2006).
[26] INTERNATIONAL ORGANIZATION FOR STANDARDIZATION, Asset Management: Overview, Principles and Terminology, ISO 55000:2024, ISO, Geneva (2024).
[27] GLOBAL FORUM ON MAINTENANCE & ASSET MANAGEMENT, The Maintenance Framework, 2nd edn (2021), https://gfmam.org/webform/download-the-maintenance-framewo.
[28] EVANS, J.R., LINDSAY, W.M., Managing for Quality and Performance Excellence, 7th edn, Thomson South-Western, Mason, OH (2008).
[29] INTERNATIONAL ORGANIZATION FOR STANDARDIZATION, Petroleum, Petrochemical and Natural Gas Industries: Collection and Exchange of Reliability and Maintenance Data for Equipment, ISO 14224:2016, ISO, Geneva (2016).
[30] WIREMAN, T., Maintenance Work Management Process, Maintenance Strategies Series, Vol. 3, Reliabilityweb.com (2013).
[31] SOCIETY FOR MAINTENANCE AND RELIABILTY PROFESSIONALS, SMPR Best Practices, 6th edn (2020), https://smrp.org/Resources/SMRP-Library/Best-Practices-Metrics-Guidelines.
[32] EUROPEAN STANDARDS, Maintenance: Maintenance Key Performance Indicators, BS EN 15341:2022, CEN, Brussels (2022).
[33] AMERICAN SOCIETY FOR TESTING MATERIALS, Standard Guide for Practical Lubricant Conditional Data Trend Analysis, ASTM D7669-20, ASTM International, West Conshohocken, PA (2020).

[34] INTERNATIONAL ORGANIZATION FOR STANDARDIZATION, Condition Monitoring and Diagnostics of Machines: Requirements for Qualification and Assessment of Personnel — Part 4: Field Lubricant Analysis, ISO 18436-4:2014, ISO, Geneva (2014).
[35] INTERNATIONAL ORGANIZATION FOR STANDARDIZATION, Mechanical Vibration — Evaluation of Machine Vibration — Evaluation of Machine Vibration by Measurement on Non-rotating Parts — Part 1: General Guidelines, ISO 20816-1:2016, ISO, Geneva (2016).
[36] STANDARDS AUSTRALIA, Test Methods for Electric Cables, Cords and Conductors: Conductors and Metallic Components, AS/NZS 1660.1:1998, SA/SNZ, Sydney (1998).
[37] INTERNATIONAL ORGANIZATION FOR STANDARDIZATION, Condition Monitoring and Diagnostics of Machines: Requirements for Qualification and Assessment of Personnel — Part 7: Thermography, ISO 18436-7:2014, ISO, Geneva (2014).
[38] EUROPEAN STANDARDS, Non-destructive Testing: Terminology - Terms Used in Ultrasonic Testing with Phased Arrays, EN 16018:2011, CEN, Brussels (2011).
[39] INTERNATIONAL ORGANIZATION FOR STANDARDIZATION, Non-destructive Testing of Welds: Ultrasonic Testing — Use of Automated Phased Array Technology, ISO 13588:2019, ISO, Geneva (2019).
[40] AMERICAN SOCIETY FOR TESTING MATERIALS, Standard Guide of Magnetic Particle Testing, ASTM E709-21, ASTM International, West Conshohocken, PA (2021).
[41] INTERNATIONAL ORGANIZATION FOR STANDARDIZATION, Non-destructive Testing: Magnetic Particle Testing — Part 1: General Principles, ISO 9934-1:2016, ISO, Geneva (2016).
[42] AMERICAN PETROLEUM INSTITUTE, Tank Inspection, Repair, Alteration and Reconstruction, API STD 653:2018, 5th edn., API, Washington (2018).
[43] INTERNATIONAL ORGANIZATION FOR STANDARDIZATION, Non-destructive Testing: Aids to Visual Inspection — Selection of Low-power Magnifiers, ISO 3058:1998, ISO, Geneva (1998).
[44] INTERNATIONAL ORGANIZATION FOR STANDARDIZATION, Non-destructive Testing of Welds: Visual Testing of Fusion-welded Joints, ISO 17637:2016, ISO, Geneva (2016).
[45] EUROPEAN STANDARDS, Non-destructive Testing — Ultrasonic Examination — Part 1: General Principles, EN 583-1:1998, CEN, Brussels (1998).
[46] EUROPEAN STANDARDS, Non-destructive Testing: Ultrasonic Thickness Measurement, EN 14127:2011, CEN, Brussels (2011).
[47] INTERNATIONAL ORGANIZATION FOR STANDARDIZATION, Non-destructive Testing: Ultrasonic Thickness Measurement, ISO 16809:2017, ISO, Geneva (2017).
[48] INTERNATIONAL ORGANIZATION FOR STANDARDIZATION, Non-destructive Testing of Welds: Ultrasonic Testing — Techniques, Testing Levels and Assessment, ISO 17640:2018, ISO, Geneva (2018).
[49] AMERICAN SOCIETY FOR TESTING MATERIALS, Standard Practice for Examination of Carbon Steel Welds Using Eddy Current Analysis, ASTM E3052-21, ASTM International, West Conshohocken, PA (2021).
[50] INTERNATIONAL ORGANIZATION FOR STANDARDIZATION, Non-destructive Testing — Equipment for Eddy Current Examination — Part 1: Instrument Characteristics and Verification, ISO 15548-1:2013, ISO, Geneva (2013).

[51] INTERNATIONAL ORGANIZATION FOR STANDARDIZATION, Non-destructive Testing: Eddy Current Testing — General Principles, ISO 15549:2019, ISO, Geneva (2019).
[52] INTERNATIONAL ORGANIZATION FOR STANDARDIZATION, Non-destructive Testing: Equipment for Eddy Current Examination — Array Probe Characteristics and Verification, ISO 20339:2017, ISO, Geneva (2017).
[53] INTERNATIONAL ORGANIZATION FOR STANDARDIZATION, Non-destructive Testing: Pulsed Eddy Current Testing of Ferromagnetic Metallic Components, ISO 20669:2017, ISO, Geneva (2017).
[54] INTERNATIONAL ORGANIZATION FOR STANDARDIZATION, Non-destructive Testing of Welds: Penetrant Testing — Acceptance Levels, ISO 23277:2015, ISO, Geneva (2015).
[55] INTERNATIONAL ORGANIZATION FOR STANDARDIZATION, Non-destructive Testing: Penetrant Testing — Part 1: General Principles, ISO 3452-1:2021, ISO, Geneva (2021).
[56] EUROPEAN STANDARDS, Non-destructive Testing: Terminology — Part 9: Terms Used in Acoustic Emission Testing, EN 1330-9:2017, CEN, Brussels (2017).
[57] EUROPEAN STANDARDS, Non-destructive Testing: Acoustic Emission Testing — General Principles, EN 13554:2011, CEN, Brussels (2011).
[58] EUROPEAN STANDARDS, Non-destructive Testing: Acoustic Emission Testing - Examination of Metallic Pressure Equipment During Proof Testing, BS EN 14584:2013, CEN, Brussels (2013).
[59] AMERICAN SOCIETY FOR TESTING MATERIALS, Standard Practice for Flux Leakage Examination of Ferromagnetic Steel Tubular Products, ASTM E507-21, ASTM International, West Conshohocken, PA (2021).
[60] AMERICAN SOCIETY FOR TESTING MATERIALS, Standard Practice for Radiographic Examination of Weldments Using Industrial X-Ray, ASTM E1032-19, ASTM International, West Conshohocken, PA (2019).
[61] INTERNATIONAL ORGANIZATION FOR STANDARDIZATION, Non-destructive Testing: Radiographic Inspection of Corrosion and Deposits in Pipes by X- and Gamma Rays — Part 1: Tangential Radiographic Inspection, ISO 20769-1:2018, ISO, Geneva (2018).
[62] INTERNATIONAL ORGANIZATION FOR STANDARDIZATION, Non-destructive Testing: Radiographic Testing of Metallic Materials Using Film and X- or Gamma Rays — Basic Rules, ISO 5579-1:2013, ISO, Geneva (2013).
[63] STANDARDS AUSTRALIA, Cranes, Hoists and Winches, Part 1: General Requirements, AS 1418.1:2021, SA, Sydney (2021).
[64] STANDARDS AUSTRALIA, Cranes, hoists and winches — Safe use, Part 15: Concrete placing equipment, AS 2550.15:2019, SA, Sydney (2019).
[65] AMERICAN CONFERENCE OF GOVERNMENTAL INDUSTRIAL HYGIENISTS, Industrial Ventilation: A Manual of Recommended Practice for Design, 31st edn., ACGIH, Ohio (2019).
[66] AMERICAN PETROLEUM INSTITUTE, Welded Tanks for Oil Storage, API STD 650:2013, 13th edn., API, Washington (2013).
[67] AMERICAN PETROLEUM INSTITUTE, Design and Construction of Large, Welded, Low Pressure Storage Tanks, API STD 620:2018, 12th edn., API, Washington (2018).
[68] STANDARDS AUSTRALIA, Inspection and Integrity Monitoring of Large Steel Vertical Petroleum Storage Tanks, AS 4971-2008, SA, Sydney (2008).
[69] ENGINEERING EQUIPMENT AND MATERIALS USERS ASSOCIATION, Above Ground Flat Bottomed Storage Tanks — A Guide to Inspection, Maintenance and Repair, Publication 159, EEMUA, London (2018).

[70] STANDARDS AUSTRALIA, Pressure Equipment — Hazard Levels, AS 4343:2014, SA, Sydney (2014).
[71] STANDARDS AUSTRALIA, Pressure equipment — In-service Inspection, AS/NZS 3788:2006, SA/SNZ, Sydney (2006).
[72] STANDARDS COUNCIL OF CANADA, Boiler, Pressure Vessel and Pressure Piping Code, CSA B51:19, CSA, Ottawa (2019).
[73] STANDARDS AUSTRALIA, Pressure Equipment, AS/NZS 1200:2015, SA/SNZ, Sydney (2015).
[74] STANDARDS AUSTRALIA, Pressure Vessels, AS 1210-2010, SA, Sydney (2010).
[75] STANDARDS AUSTRALIA, Pressure Equipment — Operation and Maintenance, AS 3873-2001, SA, Sydney (2001).
[76] AMERICAN SOCIETY OF MECHANICAL ENGINEERS, Boiler and Pressure Vessel Code Section VIII-Rules for Construction of Pressure Vessels, BPVC.VIII.1-2023, ASME, New York (2023).
[77] AMERICAN PETROLEUM INSTITUTE, Piping Inspection Code: In-service Inspection, Repair, Rating, and Alteration of Piping Systems, API 570:2016, 4th edn., API, Washington (2016).
[78] STANDARDS AUSTRALIA, Pressure piping, AS 4041-2006, SA, Sydney (2006).
[79] AMERICAN SOCIETY OF MECHANICAL ENGINEERS, Process Piping, B31.3 — 2020, ASME, New York (2020).
[80] STANDARDS AUSTRALIA, Boilers: Safety Management and Supervision Systems, AS 2593:2021, SA, Sydney (2021).
[81] STANDARDS AUSTRALIA, Safety Valves, Other Valves, Liquid Level Gauges, and Other fittings for Boilers and Unfired Pressure Vessels, AS 1271-2003, SA, Sydney (2003).
[82] BRITISH STANDARDS INSTITUTION, Code of Practice for Design and Construction of Glass-reinforced Plastics (GRP) Piping Systems for Individual Plants or Sites, BS 7159:1989, BSI, London (1989).
[83] SOUTH AFRICAN BUREAU OF STANDARDS, Glass-Reinforced Polyester (GRP) Laminates, SANS 141, Ed. 2.2, SABS, South Africa (2006).
[84] INTERNATIONAL ATOMIC ENERGY AGENCY, Computer Security for Nuclear Security, IAEA Nuclear Security Series No. 42-G, IAEA, Vienna (2021).
[85] OECD NUCLEAR ENERGY AGENCY, Managing Environmental and Health Impacts of Uranium Mining, NEA No. 7062, OECD Nulcear Energy Agenct, Issy-les-Moulineaux (2014).
[86] INTERNATIONAL ATOMIC ENERGY AGENCY, The Long Term Stabilization of Uranium Mill Tailings, IAEA-TECDOC-1403, IAEA, Vienna (2004).
[87] INTERNATIONAL ATOMIC ENERGY AGENCY, Management of Residues Containing Naturally Occurring Radioactive Material from Uranium Production and Other Activities, IAEA Safety Standards Series No. SSG-60, IAEA, Vienna (2021).
[88] INTERNATIONAL ATOMIC ENERGY AGENCY, Guidebook on Good Practice in the Management of Uranium Mining and Mill Operations and the Preparation for their Closure, IAEA-TECDOC-1059, IAEA, Vienna (1998).
[89] STANDARDS AUSTRALIA, Structural Steel Welding — Part 6: Welding Stainless Steels for Structural Purposes, AS/NZS 1554.6:2012, SA/NZS, Sydney (2012).
[90] STANDARDS AUSTRALIA, Structural Steel Welding — Part 5: Welding of Steel Structures Subject to High Levels of Fatigue Loading, AS/NZS 1554.5:2014, SA/NZS, Sydney (2014).
[91] AMERICAN PETROLEUM INSTITUTE, Risk Based Inspection, API 580:2016, 3rd edn., API, Washington (2016).

[92] AMERICAN PETROLEUM INSTITUTE, Damage Mechanisms Affecting Fixed Equipment in the Refining Industry, API RP 571:2020, API, Washington (2020).

[93] STANDARDS AUSTRALIA, Non-destructive Testing — Determination of Thickness, Part 3: Use of Ultrasonic Testing, AS 2452.3-2005, SA, Sydney (2005).

[94] AMERICAN SOCIETY OF MECHANICAL ENGINEERS, Boiler and Pressure Vessel Code Section V - Nondestructive Examination, BPVC.V-2023, ASME, New York (2023).

[95] BRITISH STANDARDS INSTITUTION, Lining of Equipment with Polymeric Materials for the Process Industries: Specification for Lining with Sheet Thermoplastics, BS 6374-1:1985, BSI, London (1985).

GLOSSARY

The definitions given below may not necessarily conform to definitions adopted elsewhere for international use.

battery limit. A defined boundary between two areas of responsibility. This may be a physical boundary on of asset (e.g. tank or pipe), a geographical boundary or a time based boundary.

downtime. The period of time in which the system, typically a piece of equipment or area of a processing facility is unavailable for use.

duty/standby arrangement. Paired devices delivering backup functionality for uninterrupted plant operations. One device being designated as 'duty', while the other device(s) is 'standby' Each device can match plant demand. Normally the 'duty' will run, but if this fails, the 'standby' device will run.

maintenance induced failure. Asset failure as a direct result of performing maintenance, including procedural errors, but often a human error induced event. Over-maintaining assets can be a source of maintenance induced failures, increasing the likelihood of asset failure.

occupied building. Buildings on a site where people are routinely working, which may be either temporary or permanently occupied. Process safety risk assessments consider buildings housing occupants and the given potential for exposure to explosion, fire and toxic hazards.

overall equipment effectiveness. Is a metric used to quantify the effective use of scheduled operating time for an asset. It excludes scheduled downtime such as for planned shutdowns.

plant. Plant comprise tangible assets held by an entity for use in the production or supply of goods or services. In uranium production industries, the set of processing equipment in its entirety to produce uranium is called a plant.

- In the context of this definition, in some State's, milling is used in the broader sense to include processing, thus 'mill' is used instead of 'plant'. Strictly speaking, milling in the context of the processing of minerals is the processing of ore to reduce its particle size, especially by crushing or grinding in a mill.

registered assets, registered equipment, registered plant. A recognized authority has to issue either of the following:

- A design registration certificate, design registration letter, design approval letter or similar confirming the design of the assets, equipment or plant (as appropriate) have been registered, approved or accepted by the authority.
- A plant registration certificate, registration letter or similar confirming the assets, equipment or plant have been registered by the authority.

Not all critical assets need to be registered with local authorities.

statutory. As laid out by law or having an obligation to meet certain laws.

work order. A document that provides all the information about maintenance tasks and outlines the schedule, resources and equipment, for completing the task.

work package. A group of related tasks, typically performed in succession, simultaneously or in parallel, as part of a maintenance project. The work package is usually a physical pack that contains all documentation required to complete the work including any permits, isolations and standard operating procedures to complete the work.

The following definitions[6] have been taken from the IAEA Nuclear Safety and Security Glossary[7] and International Standard ISO 55000:2014, Asset Management[8].

asset. An item, thing or entity that has potential or actual value to an organization. Value can be tangible or intangible, financial or non-financial, and includes consideration of risks and liabilities. It can be positive or negative at different stages of the asset life. Physical assets usually refer to equipment, inventory and properties owned by the organization. Physical assets are the opposite of intangible assets, which are non-physical assets such as leases, brands, digital assets, use rights, licences, intellectual property rights, reputation or agreements. A grouping of assets referred to as an asset system could also be considered as an asset.

maintenance. The organized activity, both administrative and technical, of keeping structures, systems and components in good operating condition, including both preventive and corrective (or repair) aspects.

predictive maintenance. Form of preventive maintenance performed continuously or at intervals governed by observed condition to monitor, diagnose or trend condition indicators of a structure, system or component; results indicate present and future functional ability or the nature of and schedule for planned maintenance.

(i) Programmes of predictive maintenance are commonly abbreviated as PdM.

preventive maintenance. Actions that detect, preclude or mitigate degradation of a functional structure, system or component to sustain or extend its useful life by controlling degradation and failures to an acceptable level.

(i) Preventive maintenance is, most commonly, periodic maintenance based on time (i.e. calendar) and/or equipment usage (i.e. run hours).

---

[6] Definitions taken from the IAEA Nuclear Safety and Security Glossary have been supplemented with additional information for this publication as indicated by (i).

[7] INTERNATIONAL ATOMIC ENERGY AGENCY, IAEA Nuclear Safety and Security Glossary, Non-serial Publications, IAEA, Vienna (2022), https://doi.org/10.61092/iaea.rrxi-t56z

[8] INTERNATIONAL ORGANIZATION FOR STANDARDIZATION, Asset Management Overview, Principles and Terminology, ISO 55000:2024, ISO, Geneva (2024) [26].

ABBREVIATIONS

ACA	asset criticality assessment
ATO	authority to operate
CCD	counter current decantation
CMMS	computerized maintenance management system
DCS	digital control system
DPI	dye penetrant inspection
FMEA	failure modes and effects analysis
HDPE	high density polyethylene
HME	heavy mobile equipment
ISO	International Organization for Standardization
MFL	magnetic flux leakage
MOC	management of change
MPI	magnetic particle inspection
MRO	maximum reasonable outcome
NDT	non-destructive testing
OEM	original equipment manufacturer
PA	phased array
PPE	personal protective equipment
PRV	pressure relief valve
PSV	pressure safety valve
RBI	risk based inspection
RCA	root cause analysis
RMS	rapid motion scanner
RPN	risk priority number
SLOFEC™	Saturated Low Frequency Eddy Current
SX	solvent extraction
UT	ultrasonic thickness
VCI	volatile corrosion inhibitor
VI	visual inspection

CONTRIBUTORS TO DRAFTING AND REVIEW

Burger, A.	Rio Tinto, Rossing Uranium, Namibia
Good, C.	International Atomic Energy Agency
Gooding, T.	Rio Tinto, Ranger Uranium, Australia
Ketel, D.	Cameco, Key Lake Uranium, Canada
Lazykina, A.	International Atomic Energy Agency
Moldovan, B.	International Atomic Energy Agency
Seini, A.	EDF Energy, United Kingdom
Tjiriange, E.	China National Nuclear Corporation, Rossing Uranium, Namibia
Tyulyubayev, Z.	Karatua LLP, Kazakhstan
Valiveti, L.	International Atomic Energy Agency

Consultants Meetings

Vienna, Austria: 3–6 December 2018, 9–13 December 2019.

IAEA
International Atomic Energy Agency

CONTACT IAEA PUBLISHING

Feedback on IAEA publications may be given via the on-line form available at:
www.iaea.org/publications/feedback

This form may also be used to report safety issues or environmental queries concerning IAEA publications.

Alternatively, contact IAEA Publishing:

Publishing Section
International Atomic Energy Agency
Vienna International Centre, PO Box 100, 1400 Vienna, Austria
Telephone: +43 1 2600 22529 or 22530
Email: sales.publications@iaea.org
www.iaea.org/publications

Priced and unpriced IAEA publications may be ordered directly from the IAEA.

ORDERING LOCALLY

Priced IAEA publications may be purchased from regional distributors and from major local booksellers.

25-00656E